Barlin Orlando Olivares Campos

Banana Production in Venezuela

Novel Solutions to Productivity and Plant Health

Springer

Barlin Orlando Olivares Campos
Biodiversity Management Research Group (GESBIO-UCO)
University of Córdoba (UCO)
Córdoba, Spain

ISSN 2366-3421 ISSN 2366-343X (electronic)
The Latin American Studies Book Series
ISBN 978-3-031-34474-9 ISBN 978-3-031-34475-6 (eBook)
https://doi.org/10.1007/978-3-031-34475-6

This Springer imprint is published by the registered company Springer Nature Switzerland AG
The registered company address is: Gewerbestrasse 11, 6330 Cham, Switzerland

To my beautiful family who always supported me in my career as a scientific professional and to all researchers and professors who dedicated their time to teaching us throughout life.

Prologue

The book that Barlin offers us represents a comprehensive approach necessary for the new approach in the management of crops. It involves the capture of appropriate agronomic information in bibliographic reference systems, selection of climatic parameters and complementary physical-chemical analysis of soil samples, measurement of growth parameters and productivity of the crop in question, as well as the judicious integration of all these elements with the use of machine learning technology, thus allowing to evaluate and quantify the effect of the different relevant variables on productivity harvests, aspects of the utmost importance for agronomic management, no longer as a technological package per crop, but integrated into the soil and climate conditions of each agroecosystem.

The content presented is the result of the research carried out by Barlin Orlando Olivares to obtain his Ph.D. in agricultural, food, forestry engineering, and sustainable rural development at the University of Córdoba (Spain). It was a quality, innovative, and interesting research process, which he developed over 5 years during different research stays in internationally renowned centers in Argentina, Chile, Costa Rica, Colombia, Panama, and Spain, first being a doctoral fellow from the Ibero-American Postgraduate University Association (Spain) and later as a scholarship holder of international mobility programs from Banco Santander, Erasmus+, Campus of International Excellence in Agrifood (ceiA3), Campus of International Excellence in Environment, Biodiversity and Global Change (CEICAMBIO), Carolina Foundation, and the Ibero-American General Secretary.

The beginnings of this investigation represented Barlin with a series of challenges in his professional and personal life. In 2018, he decided to embark on the path of doctoral training in Spain, far from his native country, which represented a great challenge for Barlin that he faced with great optimism and courage. At a professional level, he was able to participate in various pieces of training to carry out new approaches and innovative lines of research, based on the analysis of machine learning applied to the Venezuelan banana sector, which he took advantage of with skill, cunning, and enthusiasm. In general, the fruit of his efforts throughout this scientific journey was an excellent Ph.D. thesis, which he defended in January 2022, from which this book emerges.

Barlin masterfully addresses the appropriate selection and assessment of edaphic and climatic factors that increase the risk of *Fusarium oxysporum* f. sp. *cubense* (*Foc*) tropical race 4 (TR4) in banana agricultural production, banana wilting, and the prediction of productivity in two emblematic banana areas of Venezuela, all with the careful selection of advanced statistical tools and the generation of patterns with the use of algorithms and predictive models, validated with machine learning technology.

The book includes seven chapters that aim to offer an updated and practical analysis of the current state of some banana soils in Venezuela and its influence on productivity and the appearance of diseases such as Fusarium Wilt and Banana Wilt. It consists of three general chapters on the current situation of Venezuelan banana soils, methodological aspects, and a general summary of the results; and another four more specific chapters where the relevant knowledge about the lethal disease represented by Fusarium Wilt and the role of the soil in the appearance of another disease called Banana Wilt generated by a complex bacterial fungus are addressed in the first chapters (Chaps. 3 and 4), both diseases of great importance for the Venezuelan banana sector, the first (Fusarium Wilt) due to the risk of appearance and the second (Banana Wilt) due to reduced yields. Reading these chapters leads us to delve into the complex system that constitutes pathogens, the banana as a host plant, and the environmental conditions of the environment, with novel results for Venezuela, due to the use of machine learning algorithms.

Subsequently, in the other chapters (Chaps. 5 and 6), the potential that machine learning algorithms have in distinguishing productivity levels is reported, as well as demonstrating for the first time that, in the field of soil science, the variables of soil morphological characteristics such as texture, dry consistency, biological activity, reaction to HCl, and type of structure could be new promising soil indicators for banana productivity in Venezuelan soils. The study developed in Chap. 6 is a pioneer, as far as is known, in showing results of the application of categorical regression in banana soils, and a first in the environmental area of Venezuela.

Indeed, each topic has been treated with academic and scientific rigor but also serves as a basic source of knowledge for students, technicians, and professionals in plant health, soil sciences, and other professions related to agricultural sciences, as well as for all the actors that participate in the different links of the agricultural value chain, from the production to the commercialization of bananas. Barlin, together with the support of the other authors, with their deep knowledge of the subject, has managed to ensure that in each chapter the reader finds aspects of scientific and technical interest, but combined with practical aspects, and with an understandable language with the firm hope of that also serves for Latin American banana countries.

The book was born with a great opportunity, at a key moment, because of the history of *Foc* TR4 in Latin America and the Caribbean, whose recent appearance in Colombia and Peru puts Venezuela at risk. Likewise, the productivity of bananas could show significant advances with the implementation of the results presented here. Throughout the different chapters, it can be seen how the generation of new knowledge has been and is essential so that soil management strategies can be carried out on a scientific basis.

The results shown in the book reflect that the use of machine learning as a tool to program machines by different algorithms allows systems to be trained for the changing spatiotemporal situation and uneven panorama. Understanding its foundations and training and learning models, it can be used to classify or predict, generating support decisions regarding diseases in bananas, and with this classify lots of bananas with different yields based on soil variables. The content shows that the appropriate use of machine learning in the banana sector is of vital importance and relevance, and a scientific reflection regarding its applications and foundations.

With the book as a product of the editorial work, it is intended to facilitate the transfer of knowledge stored by scientists such as Barlin Olivares, so that it can reach the broadest base of professionals who carry out their activity in the banana agri-food sector in Venezuela and Latin America.

Principal Researcher
Institute for Advanced Studies (IDEA)
Caracas, Venezuela

Juan Mateus Heredia

Acknowledgments

This book is part of the work developed to obtain the title of Doctor in Agricultural, Food, Forestry Engineering, and Sustainable Rural Development from the University of Córdoba, Spain. With the preparation of this book, one of the most intense and beautiful stages of my career came to an end, hand in hand with the University of Córdoba and the Institute of Sustainable Agriculture of the Higher Council for Scientific Research (IAS-CSIC), for which I want to thank in these lines to my directors, stay supervisors, family, and friends for all the close human and academic support received during these years, without you, it would not have been possible to achieve this important achievement for me.

First, I would like to thank the Ibero-American Postgraduate University Association (AUIP) for granting me the doctoral scholarship and especially Dr. Arturo Chica from the University of Córdoba for all the support received during these years.

Secondly, to my directors Dr. José Alfonso Gómez and Dra. Blanca B. Landa from The Institute for Sustainable Agriculture (IAS) of the Spanish National Research Council (CSIC) and supervisors Dra. Deyanira Lobo and Dr. Juan Carlos Rey from the Central University of Venezuela, from whom I learned from their experience, their professional practice, and human quality, are always successful in the guidance, supervision, closeness, and support received. To Dr. Mónica Balzarini, Dr. César Acevedo, Dir. Jacob Pitti, Dr. Edilberto Montenegro, Dr. Rafael Granados, and Dr. Ximena Cifuentes, who allowed me to carry out the different research stays in Latin America. Also, to the valuable scientists such as Dr. Juan Antonio Navas, Dr. Julio Calero, Dr. Angelica Rueda, Dr. Andres Vega, and Dr. Miguel Araya-Alman for supporting the research work.

I want to acknowledge the financial support for international mobility of the Ibero-American Secretary General with Fundación Carolina and Action KA107 of Erasmus+ Program from Agri-food Campus of International Excellence (CEIA3) (2020), the scholarships of the Campus of International Excellence for the Environment, Biodiversity and Global Change (CEICambio) (2019–2021), and Santander Iberoamérica (2019). Also, the project "Technological innovations for the management and improvement of the quality and health of banana soils in Latin

America and the Caribbean" financed by FONTAGRO and coordinated by Bioversity International (before INIBAP) and project SHui (European Commission Grant Agreement number: 773903).

I also want to thank my beautiful family, who from Venezuela supported me and encouraged me to continue and overcome the nostalgia and sadness of being far from my land. To my parents Angelica and Orlando who have been my inspiration and to my sister Angherlin who has always been there when I needed her. To my aunts, uncles, cousins, and my grandmothers Barbara and Lina for their unconditional love.

Thank those who were part of this long but beautiful path in Spain; to Patricia San Segundo, Mario, and the children Virginia and Roberto for welcoming me and integrating me with all that love and sympathy; to Rosa Camacho and Luis Blanco for welcoming me into their home multiple times, as well as to all my friends and colleagues who were somehow present at the stays. Finally, to Juan Francisco Gutiérrez (Juan Fran) and his beautiful family, for making me feel at home, for supporting me emotionally, for not leaving me alone, and for reminding me that, despite being far from home, from my family, from my land; the love I receive makes up for it and fills me with happiness.

Córdoba, Spain Barlin Orlando Olivares Campos

About This Book

Banana, the edible fruit of Musaceae, is a staple food for more than 400 million people worldwide due to their nutritional and energy attributes. This makes Musaceae a crop of worldwide relevance, particularly in tropical regions, highlighting the impact of improved Musaceae cropping systems in the current efforts worldwide oriented toward a new agricultural revolution based on sustainable intensification.

Artificial intelligence (AI) in agriculture is a reality that has a lot of weight in current agricultural production. In principle, it is about implementing state-of-the-art technology to increase the productivity and sustainability of the field. Digital technological advances and agricultural research are the basis for achieving competitive campaigns in various markets. To maintain agricultural activity, a model of modern agriculture must be in place.

To achieve this, better practices for food production based on scientific and technical research capable to consider the complexity and variability within the agri-food sector are necessary. In this sense, terms that are an important part of the new agricultural scenario in the Venezuelan banana sector are addressed with the firm purpose of being a material that allows updating these important aspects to grow bananas profitably and without losses in Venezuela. The research presented in this book is oriented toward providing answers to the causes of two aspects considered of high relevance for banana production, both affecting productivity and sustainability through the application of machine learning algorithms, always addressed for the Venezuelan conditions, one of the world's largest producing countries:

1. The impact of phytosanitary risks related to Fusarium wilt and the influence of the soil on the incidence of Banana Wilt (BW) caused by a fungal-bacterial complex.
2. An observed trend toward productivity loss and soil quality decline in some commercial farms of Aragua and Trujillo states in Venezuela.

Throughout the following chapters, the different aspects of the investigation have been reflected. Chapter 1 presents the research background, with special emphasis on the overview of Venezuela's main banana production problems. Chapter 2 develops the theoretical-methodological framework adopted for this study, the specific methodology used for the studies related first to plant health and second to the influence of the soil on banana productivity, as well as the research strategies that led to the searched results. The first issue, related to banana plant health, has been covered in two consecutive studies. Firstly, in Chap. 3, an analysis of the threat of banana Fusarium wilt (*Fusarium oxysporum f.* sp. *cubense (Foc)* tropical race 4 (TR4)) in banana production systems in Venezuela is presented. This chapter synthetically characterizes reliable information on the biotic and abiotic factors related to *Foc* TR4 occurrence, in conjunction with a risk analysis and climate suitability maps for *Foc* TR4 in Venezuela. This chapter can serve as a basic summary of the available knowledge for use by plant health technicians, professionals, and other stakeholders concerning disease management.

The research oriented toward the plant health issues in bananas is completed with the study presented in Chap. 4. This chapter analyzes the relationship between soil properties and the incidence of Banana Wilt (BW), a disease of unknown etiology, that is attributed to be caused by a fungal-bacterial complex, in a case study of a commercial banana farm in the state of Aragua in Venezuela, whose incidence has reduced the planted area by more than 35.0% in recent years. The application of the Random Forest algorithm allowed the classification of the incidence of BW in the lacustrine soils of Venezuela with good precision based on the physical and chemical soil properties, being an effective tool for decision-making in the field. In addition, the use of soil information in banana areas of Venezuela allowed the identification of banana lots with high incidences of BW also using the Random Forest algorithm. The model showed that the incidence level (low or high) of Banana Wilt could be distinguished through its relationship with Zn, Fe, K, Ca, Mn, and clay content in the soil. These results can contribute to improving our understanding of the basic mechanisms and progression of BW incidence and identify soil variables that can play a determinant role in predicting the risk and evolution of BW in banana farms in tropical lacustrine soils.

The second issue, related to the relationship between banana productivity and soil properties, has been covered also in two studies. Chapter 5 contains the research oriented toward the development of an empirical correlation model to predict productivity based on soil characteristics. Five soil properties were found to have a clear agronomic and environmental importance: Mg, resistance to penetration, total microbial respiration, soil bulk density, and free-living omnivorous nematodes. This model could be used at the field level for the reliable identification of areas of high and low banana productivity in the studied areas of Venezuela.

Finally, Chap. 6 presents a study that can broaden the usefulness of soil information derived from soil profile descriptions. It validated the hypothesis that it is possible to delimit areas of different productivity within banana farms, in the two main banana-producing areas of Venezuela (Aragua and Trujillo states) using soil morphological properties (e.g., soil structure). For this, we developed a model of categorical regression prediction calibrated with soil morphological properties such as biological activity, texture, dry consistence, reaction to HCl, and structure type. In the future, if further studies are conducted validating this approach in other environmental conditions, banana productivity could be improved using information that might be already available or can be acquired at a moderate cost using standard soil profile descriptions.

This book combines a systematic bibliographic review, crop, and soil information of different farm types in Venezuela with soil profile descriptions. Using that information, it has validated the hypothesis that identifying the soil's abiotic properties, the banana plant's predisposition to the BW disease and the potential productivity of the crop can be predicted. This approach can allow the differentiation of zones with different levels of productivity and BW risk, and as an immediate consequence, avoid areas of high risk or low productivity, or adapt agronomical practices to enhance the productivity and sustainability of banana cropping systems in Venezuela.

Specific Research Goals

The general objective of the research carried out in this book was to determine and quantify the influence of soil edaphic factors and agro-environmental variables in the differences in banana productivity and susceptibility to Banana Wilt disease occurring in areas of interest for banana commercial farming in Venezuela. This general objective was oriented through these four specific objectives:

(a) To identify and analyze the main agro-environmental factors that can influence Fusarium Wilt disease caused by *Fusarium oxysporum f.* sp. *cubense (Foc)* tropical race 4 (TR4) and its dissemination in the Venezuelan production systems of Musaceae through a review of the current state of knowledge
(b) To evaluate the incidence of Banana Wilt and its relationship with the soil's physical and chemical properties in a case study in a Venezuelan banana plantation through the performance of supervised methods such as the Orthogonal Least Squares Discriminant Analysis (OPLS-DA) and the Random Forest algorithm
(c) To identify the main edaphic variables most correlated to banana productivity in Venezuela and explore the development of an empirical model to predict this productivity from numerical soil characteristics
(d) To study the potential use of soil morphological properties to differentiate levels of banana productivity in Venezuelan conditions using a prediction model based on soil categorical properties

This book represents an important contribution to the knowledge of the banana soils of Venezuela due to the current management of the banana systems in Venezuela. We are aware that the study can be improved through its systematic use in new locations, so it is our intention and hope that other international scientific community research groups join this task to produce improved versions of it.

The book also discusses a series of topics that reflect the thinking and logic of why the primary choice of soil variables that support the basis of the study. The main use of this book is intended for agricultural extension agents, agricultural service provider agents of the banana sector that assist producers of all types, and is also ideal for producers and technical personnel of large and medium-sized farms that design their production alternatives and plant health problem-solving in Latin America and the Caribbean.

Contents

About the Author

Barlin Orlando Olivares Campos He is hired as a researcher in the Biodiversity Management Research Group (GESBIO-UCO) and Completed his Ph.D. Cum Laude in Agricultural, Food, Forestry Engineering, and Sustainable Rural Development from the Agrifood Campus of International Excellence (CeiA3) of the University of Cordoba (UCO), Spain (2022) through a scholarship from the Iberoamerican Postgraduate University Association (AUIP). His outstanding scientific activity earned him the UCO Extraordinary Ph.D. Award (2022).

He is an Agricultural Engineer from the Central University of Venezuela (2009), and received a master's degree in Environmental Technology from the International University of Andalusia, Spain (2015) through a Carolina Foundation scholarship.

He started in the scientific field in 2008, when he obtained a scholarship to stay at the Water Center for Arid and Semi-arid Zones of Latin America and the Caribbean (CAZALAC) based in Chile, within the framework of the UNESCO International Hydrological Program. Later, he was a young researcher at the National Institute of Agricultural Research (INIA) of Venezuela during (September 2009 to January 2014) carrying out scientific activities and knowledge transfer to farmers. At the same time, he was an instructor professor in the agricultural climatology department of the National Experimental Polytechnic University of the Bolivarian Armed Forces. He was also a researcher at the Institute for Advanced Studies Foundation, Venezuela (IDEA), during (March-August 2014) for the scientific-technical dissemination of agricultural biotechnology, and finally a researcher at the National Institute of Meteorology and Hydrology, Venezuela (INAMEH) (January 2016 to May 2017) generating scientific production of the agroecology zoning of crops.

He is the author of an outstanding scientific production of journal articles, books, and book chapters; also, he has participated in the supervision of thesis and conference presentations and obtained various fellowships, accreditations, and recognitions as well as participation in projects for the promotion and dissemination of science in agricultural spaces. He has carried out six research stays in research institutes and universities of international prestige in Argentina, Chile, Costa Rica, Colombia, Spain, and Panama.

About the Author

Chapter 1
Research Background

Abstract In terms of the gross value of their production, bananas are the fourth most important food crop in the world, after rice, wheat, and maize. Latin America dominates the world banana economy, which is grown mostly in large monoculture plantations. This chapter presents the conceptual framework of the problems that characterize most banana production systems in Venezuela. First, a brief overview of the importance of bananas in the world's food chain is presented; secondly, the overview of banana production in Venezuela is analyzed; thirdly, the main phytosanitary problems in Venezuela are mentioned; fourthly, emphasis is placed on the problems of banana soils in Venezuela; and finally, some advances on innovative solutions using machine learning are pointed out. Monitoring the soil quality of bananas and its relationship with susceptibility to diseases such as Fusarium wilt and banana wilt caused by a fungal-bacterial complex would allow banana growers to characterize the most important critical factors of their soil and identify best practices and management to avoid deterioration or to recover the lost condition as far as possible. The description of the current state of soil quality and its evolution constitutes the information base necessary to understand the causes and dynamics of the deterioration process and to design alternative technological innovations through machine learning.

Keywords Agronomy · Banana · Productivity · Soil quality · Soil fungus · Machine learning

1.1 Introduction

The term "artificial intelligence (AI)" refers to a combination of algorithms that make a machine more human-like in the way it functions (Jha et al. 2019; Bannerjee et al. 2018). There are multiple variants in this field, from machines that learn, solve, and make decisions to others that focus on emulating how people act (Jung et al. 2021). AI can be the solution so that a growing population and an increasingly

B. O. Olivares Campos, *Banana Production in Venezuela*, The Latin American Studies Book Series, https://doi.org/10.1007/978-3-031-34475-6_1

uncertain climate do not prevent us from getting enough food while at the same time protecting biodiversity and the great variety of crops.

Some of the main aspects that are being worked on to apply the possibilities of artificial intelligence in agriculture are the following: The ability to predict and evaluate soils and/or crops. Through a system of algorithms capable of concluding data, a prediction and/or climatological suitability system can be applied to agriculture that will allow all kinds of measures to be adopted. If AI is applied to meteorology, not only can rains, droughts, and other types of phenomena be anticipated, but potentially suitable areas for the appearance of some pests and plant diseases can also be established.

Through solutions with machine learning algorithms, crop yields can be improved through the management of the information obtained in the agri-food system. The use of sensors, which can be placed close to the crops or even at considerable distances such as from a satellite, will collect data and images that will guarantee the ability to react to pests, weeds, or fertilizer needs. In addition, these products can be adapted through the analysis of samples to the specific needs of the soil, varying and adjusting the formulas according to their conditions. Monitoring through AI will also make it possible to make observations and issue analyses over long periods, greatly facilitating life for the agri-food sector.

Bananas and plantains are essential crops worldwide for their economic importance and deep cultural roots. At the beginning of this study, the main problems in banana production systems in Venezuela have been identified. In this book, only the agronomic problems of plant health and banana productivity are described, focusing on the effect of the soil through the application of machine learning algorithms, mainly because one of the main problems in banana production in Venezuela is the attack of diseases and the deterioration of soil quality, for which these three main aspects were identified.

The first aspect is related to the impact of the phytosanitary risks of Fusarium wilt, where all banana-producing countries are seriously threatened worldwide by an "epidemic" caused by the fungus *Fusarium oxysporum* f. sp. *cubense* (*Foc*) Tropical Race 4 (TR4), which can generate the total loss of the harvest, so there are no effective control measures so far. The second aspect is represented by the incidence of banana wilt (BW); this is a disease of unknown etiology, which is attributed to being caused by a fungal-bacterial complex. In the case of Venezuela, the incidence has reduced the planted area by more than 35.0% in the state of Aragua. And finally, the third aspect is related to the problems of banana soils in Venezuela; the situation characterized by the application of conventional agricultural techniques and high-cost inputs has not been enough to reverse the trend toward reduced yields. This has also been related to the worsening of soil properties in banana plantations, hence the need to incorporate new study approaches with novel techniques (Fig. 1.1).

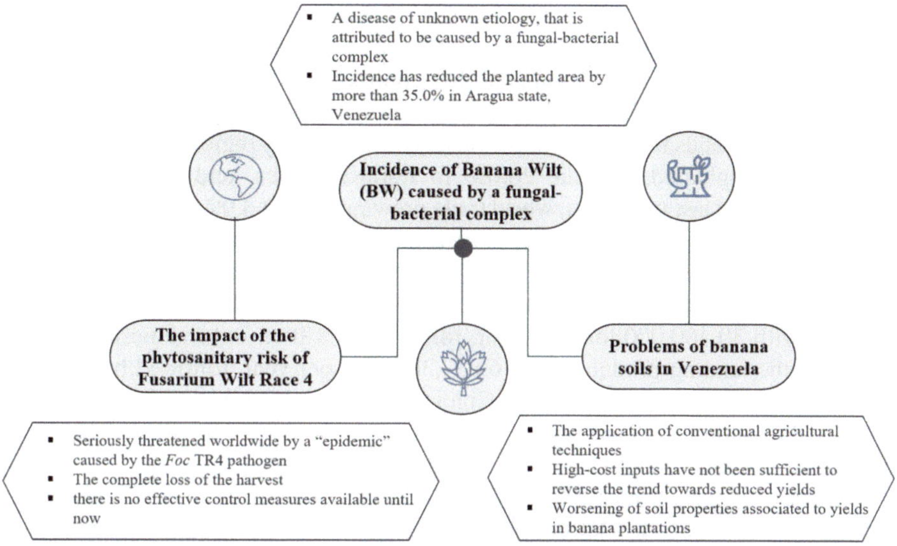

Fig. 1.1 Overview of the main banana production problems in Venezuela

1.2 Importance of Bananas in the World's Food Chain

Banana is the most consumed fruit in the world, and it is one of the most dynamic crops in international trade, being one of the most exported fruits. This dominance can be explained by the characteristics of banana plants that give them comparative advantages over other fruits. Among the most important characteristics of this crop are that banana plants reproduce asexually sprouting shoots from an underground stem (Iskandar et al. 2018), it has vigorous vegetative growth, shoots can produce a mature bunch in less than a year, the fruits can be harvested throughout the year (Scott 2021), and shoots continue to sprout from a single plant year after year, making bananas a perennial crop whose duration of plantation can last from 6 to 15 years (Varma and Bebber 2019).

In other words, Musaceae (bananas and plantains) are perennial crops that grow quickly and can be harvested throughout the year. Around 116,781,658 tons of bananas are produced in the world per year (FAO 2021a). These figures are an approximation, since most of the world's banana production, almost 60.0%, comes from relatively small plots and home gardens from which there are no reliable statistics. In many developing countries, most of the banana production is destined for self-consumption or traded locally, thus playing an essential role in food security in those countries (Coltro and Karaski 2019).

That is why the importance of bananas as a food crop in the tropics cannot be underestimated. In Africa, particularly in countries such as Uganda, for example, annual per capita consumption reaches about 243 kg, and in Rwanda, Gabon, and Cameroon, it ranges between 100 and 200 kg. In these four countries, bananas represent between 12.0 and 27.0% of the daily calorie intake of their populations (Sabiiti et al. 2018).

The FAO reports (2021b) indicate that approximately 99.0% of banana world production takes place in developing countries, and practically, all this world production corresponds to the Cavendish variety. Examined by region, 56.0% of banana production originated in Asia, 26.0% in Latin America, and 15.0% in Africa (Fig. 1.2). In the last two years, world trade grew rapidly, at 3.0% per year, and more than three-quarters of exports went to consumers in developed countries. However, Latin America and the Caribbean (LAC) is the third region in terms of banana production, behind Asia and Africa. The world's leading producing country of bananas is India with 30,460,000 tons per year, followed by China and Indonesia. In LAC, Brazil is the leading producer with 6,637,308 tons per year, followed by Ecuador, Guatemala, Costa Rica, Mexico, Colombia, Peru, Dominican Republic, and Venezuela (Table 1.1). These eight countries together account for more than half of the world's production of Cavendish bananas (Fig. 1.2).

As such, exports from LAC, the world's leading exporting region, declined by about 3% in 2021, to a total of 15.9 million tons—some 500,000 tons lower than their 2020 level. Ecuador, the largest exporter of bananas globally, registered a 3.2% decline in shipments, to approximately 6.8 million tons. Industry sources reported that, in addition to the above-mentioned obstacles, the banana industry in Ecuador

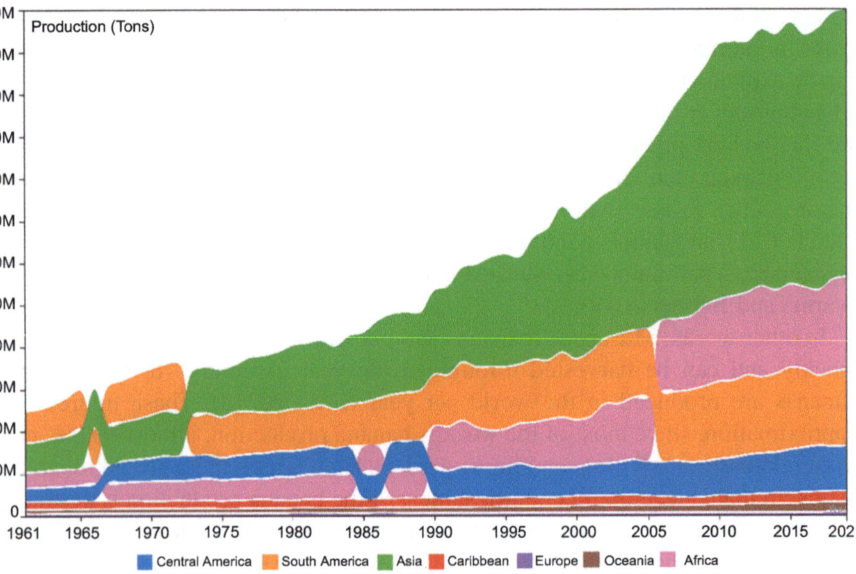

Fig. 1.2 Bump chart of historical evolution production (tons) (1961–2019) of banana cultivation in Central America, South America, Asia, the Caribbean, Europe, Oceania, and Africa. Note: Interpreting a bump chart, when a line crosses another line, that is indicative of a change in rank. In other words, a crisscross in a bump chart indicates one entity (continent) has surpassed others in absolute terms even when the comparison is based on relative ranks

Table 1.1 Banana production (tons), harvested area (ha), and yield (tons/ha) in LAC according to the FAO (2021a)

Country	Production (tons)	Harvested area (ha)	Yield (tons/ha)
Brazil	6,637,308	455,004	14.58
Ecuador	6,023,390	160,630	37.49
Guatemala	4,476,680	93,554	47.85
Costa Rica	2,528,721	53,975	46.84
Mexico	2,464,171	79,747	30.89
Colombia	2,434,900	99,417	24.49
Peru	2,314,514	160,651	14.40
Dominican Republic	1,232,039	29,111	42.32
Venezuela	884,207	50,213	17.60
Honduras	585,929	11,144	52.57
Bolivia	301,162	20,010	15.05
Panama	285,188	7792	36.60
Haiti	271,861	55,936	4.86
Cuba	263,405	36,095	7.29
Argentina	176,436	8415	20.96
Nicaragua	102,628	1824	56.26
Paraguay	94,486	8852	10.67
Belize	92,635	3044	30.43
Puerto Rico	77,123	1560	49.43
Jamaica	62,257	8239	7.55
VCT	61,001	6167	9.89
Suriname	48,461	1676	28.91
Dominica	21,262	2781	7.64
Guyana	17,205	571	30.13
El Salvador	14,028	1033	13.57
Bahamas	9928	437	22.71
Santa Lucía	9651	213	45.30
Trinidad and Tobago	3429	1020	3.36
USA	2935	327	8.97
Granada	1969	580	3.39
Barbados	995	174	5.71
Antigua and Barbuda	5	11	0.45

Note: VCT = St. Vincent and the Grenadines

was affected by higher expenditures stemming from the necessity to maintain stringent *Foc* TR4 mitigation measures given the outbreaks of *Foc* TR4 in neighboring Colombia and Peru (FAO 2021b).

The world's banana production has increased steadily between 1985 and 2019. Annual production increased by 49.0%, from 42.5 million tons in 1985–1987 to 63.4 million tons in 1998–2000 (Fig. 1.2). This increase was due, firstly, to the expansion of the cultivated area (Fig. 1.3) and, to a lesser extent, to the increase in

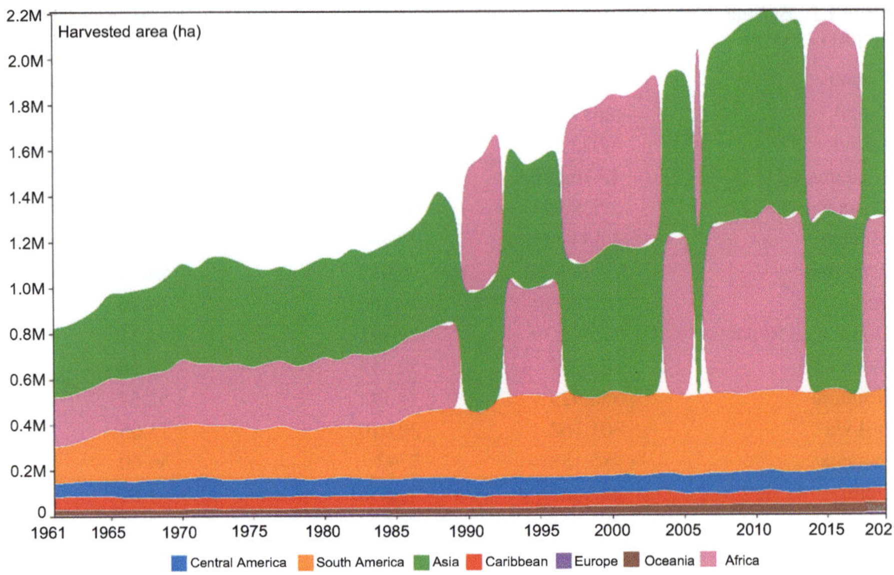

Fig. 1.3 Bump chart of historical evolution harvested area (ha) (1961–2019) of banana cultivation in Central America, South America, Asia, the Caribbean, Europe, Oceania, and Africa. Note: Interpreting a bump chart, when a line crosses another line, that is indicative of a change in rank. In other words, a crisscross in a bump chart indicates one entity (continent) has surpassed others in absolute terms even when the comparison is based on relative ranks

productivity. During this period, the average yield increased from 13.7 to 15.8 tons per hectare (15.0%) (Fig. 1.4).

According to the FAO (2021b), the projections suggest that world banana production will keep growing at a rate of 1.5% per year, reaching 135 million tons in 2028. Of all fruits, bananas would register the slowest growth since their demand is saturated in most regions and so its growth is mainly driven by population growth. Banana is the main crop in terms of volume among tropical fruits and is expected to represent approximately 53.0% of total world tropical fruit production in 2028, compared to approximately 58.0% in 2009, as demand for other tropical fruits, particularly mango and avocado, has increased rapidly.

It is expected that Asia will maintain a high share of world production at approximately 54.0%, and India will continue to be the world's largest producer of bananas with an expected volume of 33 million tons a year. Production growth in India will be supported by strong domestic demand because of higher population growth. Production in the prominent exporting region of Latin America and the Caribbean (primarily Ecuador, Brazil, Guatemala, Colombia, Costa Rica, and Mexico) is expected to reach 34 million tons, fueled by demand for imports from key customers in the markets from developed countries.

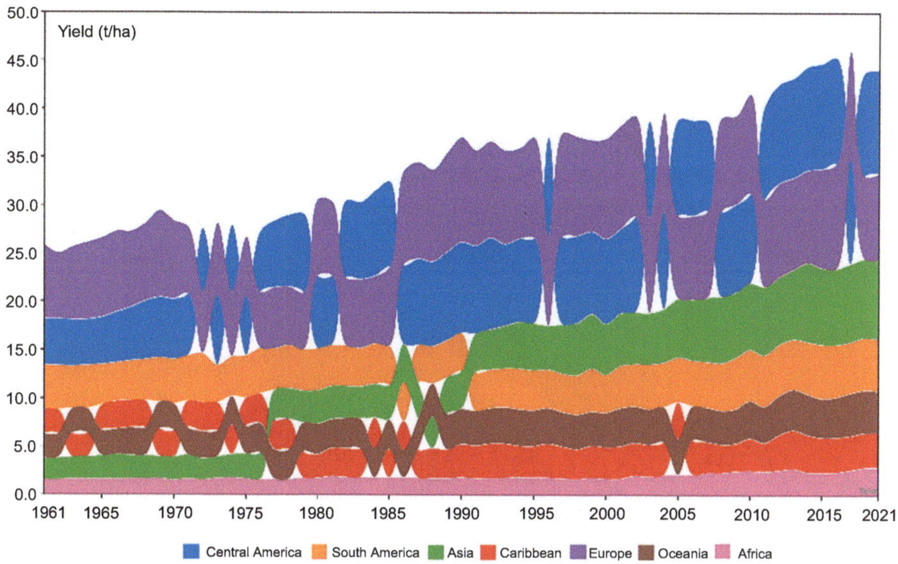

Fig. 1.4 Bump chart of historical evolution yield (tons/ha) (1961–2019) of banana cultivation in Central America, South America, Asia, the Caribbean, Europe, Oceania, and Africa. Note: Interpreting a bump chart, when a line crosses another line, that is indicative of a change in rank. In other words, a crisscross in a bump chart indicates one entity (continent) has surpassed others in absolute terms even when the comparison is based on relative ranks

In 2018–2020, Colombia ranked as the fourth largest exporter in the world with a total volume of 1.7 million tons, equivalent to approximately 9.0% of the world's total (Fig. 1.5a). Asia, the world's second-largest banana exporter, accounted for approximately 20.0% of global shipments in 2018–2020, almost entirely supplied by the Philippines, the region's leading exporter. Of all LAC, Colombia and Peru deserve special attention due to the appearance of Foc TR4. Among the main importers, the European Union and the United States of America absorbed, respectively, 32.0 and 23.0% of the world supply in 2018–2020 (Fig. 1.5b). The Russian Federation, China, and Japan are also major importers, albeit with single-digit market shares in the same year.

Given the popularity of bananas in import markets, their global value chains have been characterized by intense competition between market players at all levels, including the retail sector. This has put downward pressure on prices at each stage of the chain, producing small fluctuations in producer prices and generally staying at very low levels (FAO 2019). Combined with rising production costs, low prices and very tight profit margins make it extremely difficult to adequately compensate workers and small banana producers and are the main obstacles that producers face in the face of emerging challenges, including the imminent threat of the *Foc* TR4.

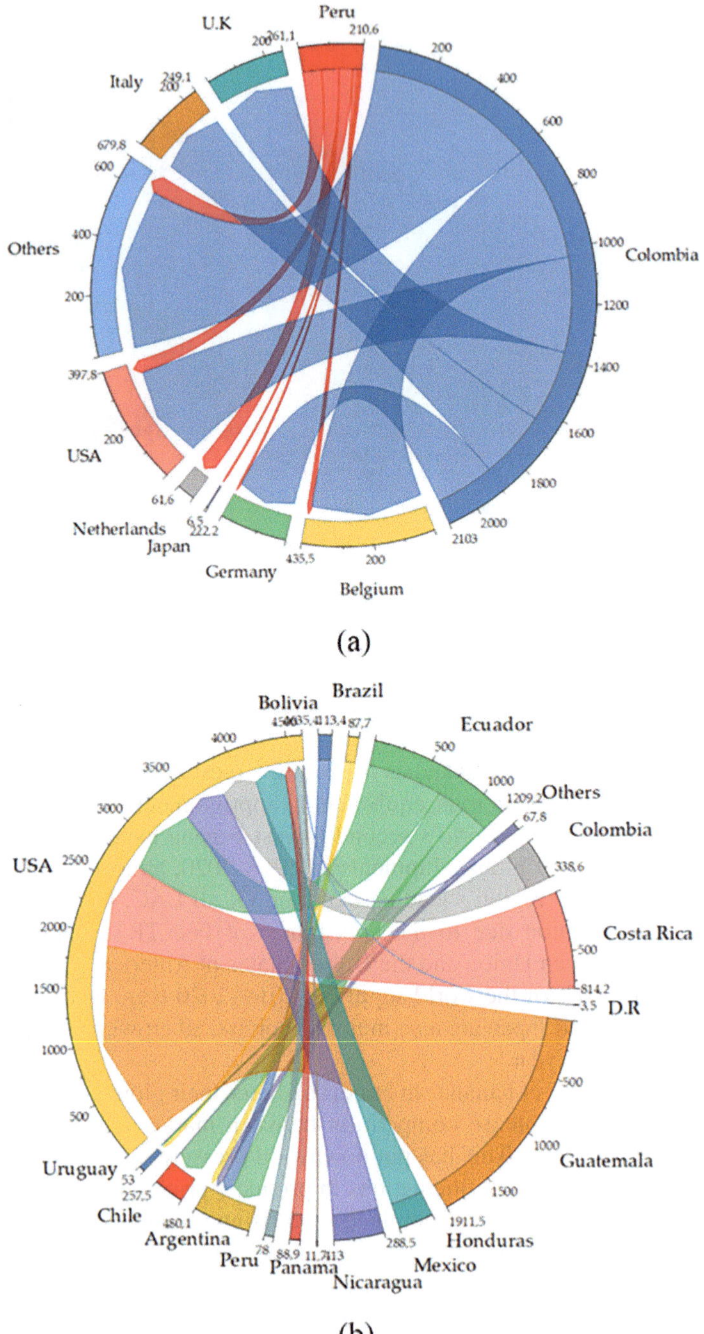

Fig.1.5 (**a**) Chord diagram of banana exports from Colombia and Peru, countries with the presence of *Foc* TR4, and (**b**) chord diagram of banana imports from Argentina, Chile, Uruguay, and the USA from LAC countries according to the data source from the FAO (2022)

1.3 Overview of Banana Production in Venezuela

Bananas represent a particularly important crop for Venezuela's economy, being one of the few alternatives to an oil producer economy. The monoculture areas in the main producing areas use banana clones (*Musa* AAA, Cavendish subgroup). Generally, these areas are characterized by being located at an altitude, not above 600 meters above sea level (masl) (Martínez et al. 2020a), concentrated mainly in the South of Lake Maracaibo (Zulia state), the Central region, the Central West, and the East regions (Fig. 1.6). However, it can be indicated that in recent years, the introduction of the tetraploid FHIA-17 (*Musa* AAAA) has been observed in the banana cultivated areas of Aragua state (Rey et al. 2020).

Production data for Venezuela between 2007 and 2018 (FAO 2021b) indicate that it ranks nine among the ten important banana-producing countries in America with an average of 480,103 tons/year (Table 1.1). For the year 2021, the production of bananas in the country was 884,207 tons (Fig. 1.7b) in a harvested area of 50,213 ha (Fig. 1.7a) with a yield of 17,60 tons/ha (Fig. 1.7c) (Table 1.1).

As can be seen in the data in Fig. 1.7a, in 1992, the harvested area reached a maximum of 58,745 ha; from that date, it suffered a sharp drop until it reached a

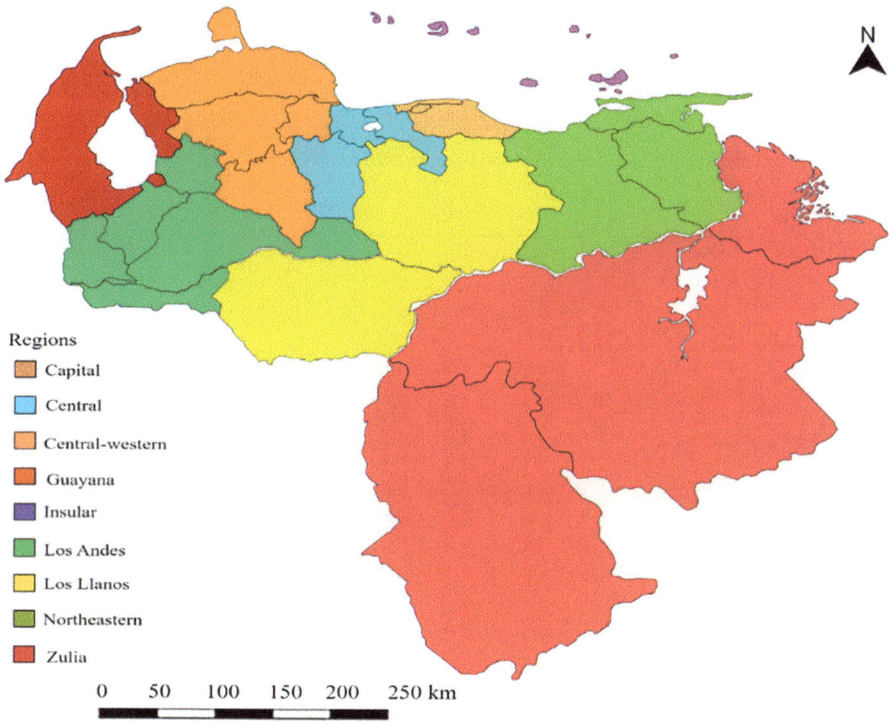

Fig. 1.6 Map of Venezuela's states, colored according to the administrative regions

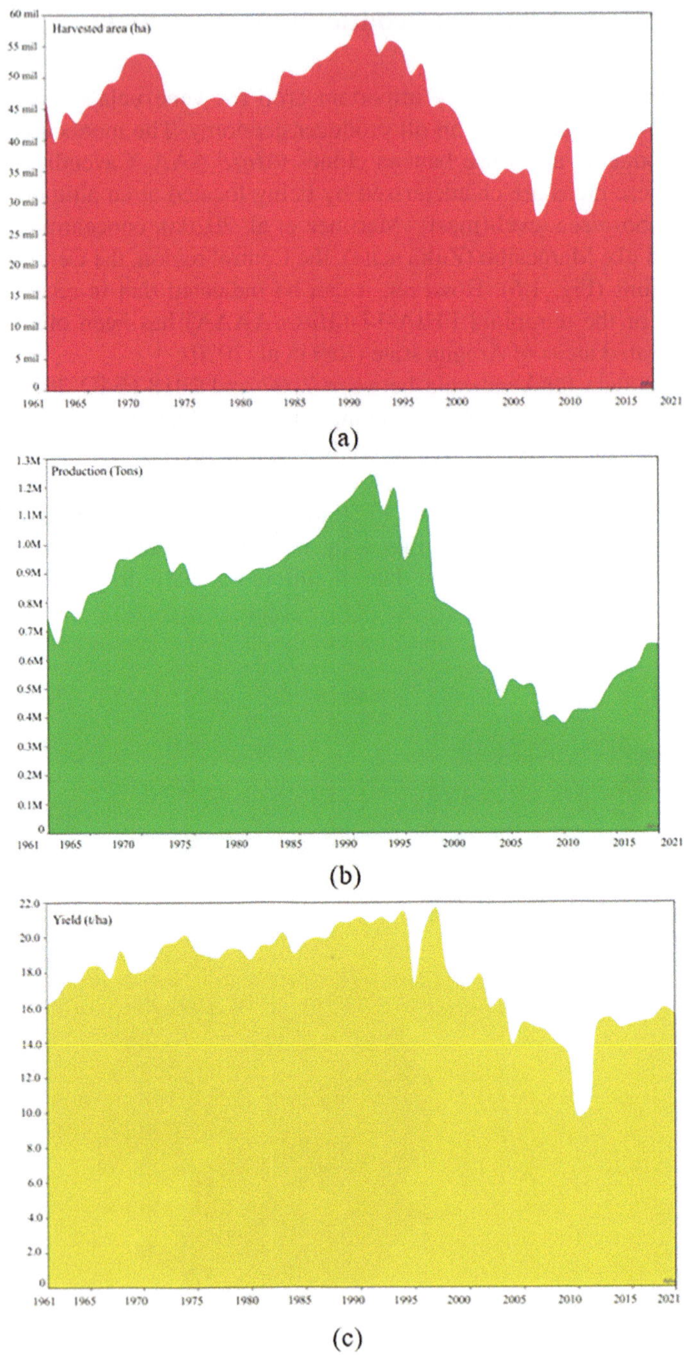

Fig. 1.7 Historical evolution (1961–2019) of banana cultivation in Venezuela. (**a**) Harvested area (ha), (**b**) production (tons), and (**c**) yield (tons/ha) according to the data source from FAOSTAT (2020)

minimum of 27,468 ha for 2008 (FAO 2021a), basically due to bad survey of primary information and the social and political conditions of the country at that time.

During the period 1961–1991, the banana national average production was 928,191 tons per year. In 1992, production reached its maximum level of 1,239,480 tons, and in 2010, its lowest point was at 75,118 tons (Fig. 1.7b). According to the FAO, the volume of production registered shows a growing trend from 2010 to 2019. On the other hand, the highest yield of national banana production was in 1997 with 21, 61 tons/ha until reaching a minimum in 2010 of 9,71 tons/ha (Fig. 1.7c). In general, there is a decline in the banana production and harvested area in Venezuela from 1992 to approximately 2010, and as of this date, a partial recovery has been noted.

Within the productive process of banana cultivation, in recent years, a deterioration of the physical, chemical, and biological factors of the soil was evidenced; as a consequence of the indiscriminate, constant, and accumulated use of agricultural products (fertilizers, pesticides, and others), required in a high proportion by intensive production systems and/or destined for export, to generate an increase in production and fruit quality. This behavior caused, in many Latin American countries, including Venezuela, a considerable decrease in production, approximately 50% from 1992 to 1994.

1.4 Main Phytosanitary Problems in Venezuela

During the last 25 years, banana production in Venezuela has faced severe challenges due mainly to the shortage of agro-inputs (seeds, fertilizers, agrochemicals), problems of access to foreign exchange to satisfy internal demand, and the inadequate management of agricultural policies, as well as the impact due to drought, pests, and diseases (Rey et al. 2020).

A recent study developed by Martínez et al. (2020a) establishes that among the main limiting factors of banana production in Venezuela are the occurrence of meteorological droughts (Olivares et al. 2017; Cortez et al. 2018), the attack of the black banana weevil (*Cosmopolites sordidus* Germar) (Rey et al. 2006, 2009), and the severe effects of diseases such as the Sigatoka complex (*Mycosphaerella* spp.), the bacteria *Erwinia carotovora*, a fungal-bacterial complex named banana wilt (BW) (Olivares et al. 2022a), Moko (*Ralstonia solanacearum*), and Fusarium wilt (*Fusarium oxysporum* f. sp. *cubense*) (*Foc*) race 1 (Fig. 1.7).

According to Martínez et al. (2020a), there is a wide differential margin between the phytosanitary limitations caused by the Black Sigatoka complex (*Mycosphaerella fijiensis* Morelet) (Hernández et al. 2006; Olivares et al. 2022b) and Yellow Sigatoka (*Mycosphaerella musicola* Leach et Mulder) and the rest of the diseases (Gómez et al. 2012). The distribution of these phytosanitary problems in regions of production of Musaceae in Venezuela is indicated in Fig. 1.8. It is observed that in the South region of Lake Maracaibo, Black Sigatoka occurs in a higher proportion concerning the rest of the diseases, followed by the cases reported for *Erwinia*

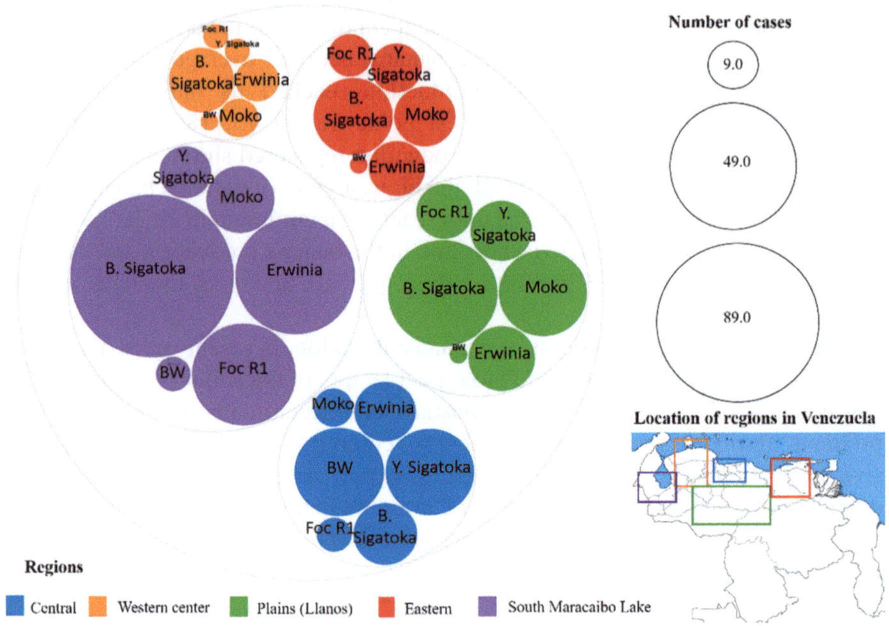

Fig. 1.8 Distribution of the main phytosanitary problems in the Musaceae production regions in Venezuela according to Martínez et al. (2020a). B. Sigatoka = Black Sigatoka (*Mycosphaerella fijiensis* Morelet); Y. Sigatoka = Yellow Sigatoka (*Mycosphaerella musicola* Leach et Mulder); Foc: *Fusarium oxysporum* f. sp. *cubense* (race 1). Erwinia = *Erwinia carotovora*; BW = fungal-bacterial complex called banana wilt; Moko (*Ralstonia solanacearum*) (Central Region, $n = 40$; Western Center Region, $n = 15$; Los Llanos Region, $n = 50$; Eastern Region, $n = 23$; South Region of Lake Maracaibo, $n = 101$)

carotovora and *Foc* race 1, considering the rest of them of less importance. Black Sigatoka appears in the rest of the regions, except in the Central one, as the main phytopathological problem.

In the Los Llanos region, Black Sigatoka occurs with very narrow differential ranges concerning the number of cases reported for Moko and Yellow Sigatoka. In the Central region, unlike the rest of the regions, Yellow Sigatoka together with a new disease called the fungal-bacterial complex (BW) represents the main disease in crops, even when Black Sigatoka is present, but its degree of attack is attenuated due to existing environmental conditions.

In those producing areas characterized by high rainfall and high temperatures, and certain drainage limitations, there may be conditioning factors for the eventual incidence and severity of diseases, which is why it is recommended to implement integrated control with the use of propagation material. Healthy (Hamed et al. 2018), restriction of movement between zones (Martínez et al. 2020b), the establishment of preventive strategies against emerging diseases to mitigate the risk of introduction and spread (Dita et al. 2018), and diversification of the production with clones resistant to new threats; disseminate this knowledge (Ahmad et al. 2020).

Biotic stress caused by emerging diseases continues to affect these production systems in various regions, causing partial damage or destruction of the fields and representing a considerable threat to the food security of Venezuela (Olivares et al. 2021a). Unfortunately, the banana is seriously threatened worldwide by a "pandemic" caused by the *Foc* TR4 fungus, which, once established in the field, can cause the complete loss of the harvest and for which there are no effective control measures until now.

The fact that the fungus is found in the soil and the pathogen can survive on it for more than 40 years complicates its control and eradication. The only viable solution today is through prevention and training of producers and monitoring of plantations to detect infected plants as early as possible and contain the spread of the fungus (Dita et al. 2017). A collaborative effort is underway between official institutions and the private sector so that this disease does not spread to countries such as Brazil, Ecuador, Costa Rica, and Venezuela (Ghag et al. 2015). At the same time, research is continuing in the search for alternatives for effective biological or chemical treatments to protect plants or reduce disease severity.

The use of varieties with resistance to the disease is a key management strategy to face the problem but supposes a great challenge: 80.0% of the world's banana production comes from varieties susceptible to this disease. Furthermore, almost 50.0% of the bananas that are produced and 95.0% of what is exported originate from a Cavendish type or variety, which is susceptible to this TR4 (Ploetz 2015). To stop the disease caused by *Foc* TR4, an integrated approach is needed, where all the actors of the system are under alert and contributing from their possibilities.

1.5 Problems of Banana Soils in Venezuela

Most of the banana plantations in Venezuela are characterized by being aged (>15 years). This aging has been identified since the end of the twentieth century, and in recent years, the trend has not improved. They are generally aged plantations with low-quality materials, a low technological level, and serious phytosanitary problems, managed by small and medium producers for subsistence purposes and, to a lesser extent, for the national market (Nava 1997; Delgado et al. 2008).

According to the studies of Rey et al. (2006), Delgado et al. (2008), Rey et al. (2009), and Delgado et al. (2010a), the application of conventional agricultural techniques and high-cost inputs has not been sufficient to reverse the trend toward reduced yields. This has also been linked to the worsening of soil properties in banana plantations (Rey et al. 2020). For example, in the last decade, the overall productivity of banana plantations in Costa Rica and Panama has shown considerable oscillations (FAO 2021a) contrary to the stability in productivity in organically managed banana farms, which are characterized by the limited use of agricultural inputs (Serrano 2003). Likewise, these organic production systems promote greater biological activity and therefore greater suppressiveness of pathogens associated with the root system than conventionally managed soils (Pocasangre et al. 2000;

Menéses et al. 2003). This results in an economic advantage for the farmer, as well as an improvement in soil quality (Rosales et al. 2008).

Nowadays, the methods used to determine the productive potential of banana soil are based on the appraisal of its physical and chemical properties (rarely considering biological properties) and on the effect of local topography and climate and agronomic management. In general, these methods are not sufficient to capture the complex interactions of the soil and its rhizosphere on banana productivity (Pattison et al. 2014). Due to this, there is a limited number of studies aimed at a broader evaluation of soil quality and its impact on banana productivity (e.g., Delgado et al. 2010a). Despite this, there is a need for further studies to enhance the understanding of soil's physical, chemical, and biological properties, quantitative or categorical, and its influence on the productivity of bananas in many countries, such as Venezuela. This is a necessary step to address and solve the problem of depletion and low productivity of many banana plantations, which has been linked to the adverse impact of the conventional production system (Gauggel et al. 2003; Pattison et al. 2004; Delgado et al. 2010b).

Recent studies have demonstrated the possibility of using soil morphological properties in studies of soil quality and plant productivity through statistical techniques and methods such as optimal scaling, categorical regression, or the analysis of categorical principal components (Calero et al. 2018; Bouma 2020), although to our knowledge, this approach has never been tried in banana farms.

1.6 Innovative Solutions Using Machine Learning

In the agricultural sector, modeling techniques are crucial for predicting crop growth and yield, allowing for more efficient and accurate decision-making now and in the future (Jayasinghe et al. 2022). Machine learning is used in the field of artificial intelligence to train predictive models that are used in different areas of science (Baviera 2017). Machine learning is a branch of artificial intelligence, where the computer increases its knowledge to fulfill a task. Through algorithms, it could identify patterns in massive data to make predictions, allowing them to be autonomous, without the need to be programmed (Wang et al. 2020; Alimi et al. 2021).

Machine learning application uses supervised (algorithms that "learn" from data entered by a person) and unsupervised (algorithms that learn from data with unlabeled elements looking for patterns or relationships between them) methods to support data analysis and calibration procedures. It consists of two phases: a training phase and a test phase; in the training phase, there is a set of data (usually between 60% and 70% of the total available data), and 40% or 30% of the data remaining is used to make predictions and to validate the performance of the algorithm (Van Engelen and Hoos 2020). Machine learning allows both identifying patterns among a considerable amount of data that may be of different natures and predicting behaviors through algorithms capable of learning and evolving based on their own experience (Ma et al. 2017; Ye et al. 2020).

There are several examples of different techniques for machine learning that might be useful for predicting the incidence of diseases in bananas and achieving a more precise diagnosis. For instance:

(a) Random forest (RF) used by Owomugisha et al. (2014), Sangeetha et al. (2020), Ye et al. (2020), Gómez-Selvaraj et al. (2020), and Olivares et al. (2021b).
(b) Support vector machines with the linear or radial kernel (SVM) used by Hou et al. (2015), Aruraj et al. (2019), and Ye et al. (2020).
(c) Classification and regression trees (CART) applied by Manjunath et al. (2019).
(d) Decision tree algorithm (C5.0) applied by Owomugisha et al. (2014).
(e) Linear discriminant analysis (LDA) developed by Companioni et al. (2005) and Olivares et al. (2022c).
(f) Artificial neural network used by Ye et al. (2020).

Many efforts have been made to calculate parameters such as vegetation indices, crop height, crop yield, leaf area index, surface soil properties, soil biomass, water stress, canopy height models, and more. A brief description of each algorithm is specified in the following paragraphs:

(a) Random forest (RF): An ensemble of many independent individual classification and regression trees (CART) and defined as Eq. 1, where h represents the RF classifier, x is the input variable, and {θk} represents the independent identically distributed random predictor variables, which are used for generating each CART (Breiman 2001; Liaw and Wiener 2002). The final response of the RF is calculated based on the output of all decision trees (Eq. 1.1).

$$\left\{h\left(x,\theta_k\right),k=1,2,\ldots i\ldots\right\} \tag{1.1}$$

(b) Linear discriminant analysis (LDA): A supervised classification method that is used to create machine learning models. These models based on dimensionality reduction are used in the detection of plant diseases (Xanthopoulos et al. 2013). Using Bayes' theorem, LDA estimates the probability that an observation, given a specific value of the predictors, belongs to each of the classes of the variable (Eq. 1.2). Finally, the observation is assigned to class k for which the predicted probability is higher.

$$P\left(\mathrm{Y}=k|\mathrm{X}=x\right) \tag{1.2}$$

(c) Classification and regression trees (CART): The representation used for CART is a binary tree. Predictions are made with CART by traversing the binary tree given a new input record. The tree is learned using a so-called "greedy" algorithm on the training data to pick splits in the tree. Stopping criteria define how much a tree learns, and pruning can be used to improve a learned tree (Quinlan 2007; Breiman et al. 1984).
(d) Radial basis function kernel Support Vector Machines (RSVM): A non-parametric supervised statistical learning classifier. The higher performance of the SVM

classifier makes it a favored alternative for plant disease detection. SVM considers the structural risk minimization (SRM) principle to maximize the margin of class separation for better generalization performance of SVM. (Vapnik 1995; Hearst et al. 1998). Two parameters need to be set when using an SVM classifier with the radial basis function kernel, i.e., the cost function (C) and the kernel width parameter (γ). The C parameter trades off the misclassification of training examples against the simplicity of the decision surface. The γ affects the smoothness of the class-dividing hyperplane (Ye et al. 2020). Eq. 1.3 shows the mathematical definition of radial basis kernel, where x and x' are vector points in any fixed dimensional space.

$$K\left(x,x'\right)=\exp\left(-\frac{x-x'^2}{2\sigma^2}\right) \tag{1.3}$$

(e) Linear kernel Support Vector Machines (LSVM): A parametric model (Cortes and Vapnik 1995). Given a set of samples xi (i = 1,2,…M), where M is the number of samples. The set has two classes; those are positive class and negative class. We denote yi = 1 for the positive class and yi = −1 for the negative class, respectively. It is possible to find a hyperplane $f(x) = 0$ that classifies the given dataset (Eq. 1.4), where w is an M-dimensional vector and b is a scalar, and they are used to define the hyperplane (Lei 2017).

$$f\left(x\right)=w^{\mathrm{T}}x+b=\sum_{j=1}^{M}w_jx_j+b=0 \tag{1.4}$$

(f) C5.0 decision tree algorithm: The main two modes for this model are a basic tree-based model and a rule-based model (Quinlan 2014). C5.0 can create an initial tree model and then decompose the tree structure into a set of mutually exclusive rules. These rules can then be pruned and modified into a smaller set of potentially overlapping rules (Kuhn et al. 2018). C5.0 uses the concept of entropy for measuring purity. The entropy of a sample of data indicates how mixed the class values are; the minimum value of 0 indicates that the sample is completely homogenous, while 1 indicates the maximum amount of disorder. The definition of entropy can be specified in Eq. 1.5; for a given segment of data (S), the term "c" refers to the number of different class levels, and pi refers to the proportion of values falling into class level I (Eq. 1.5).

$$\text{Entropy}\left(S\right)=\sum_{I=1}^{C}-p_i\log_2(p_i) \tag{1.5}$$

To quantitatively analyze the research works related to machine learning in bananas through the mathematical method known as bibliometrics, Figure 1.9a provides an overview of the current use of these machine learning techniques for fruit crops using Scopus (2000–2020) by using VOSviewer (Van Eck and Waltman 2010). The words machine learning, learning system, and fruit are the central nodes, which establishes that they are high-frequency keywords in the studies

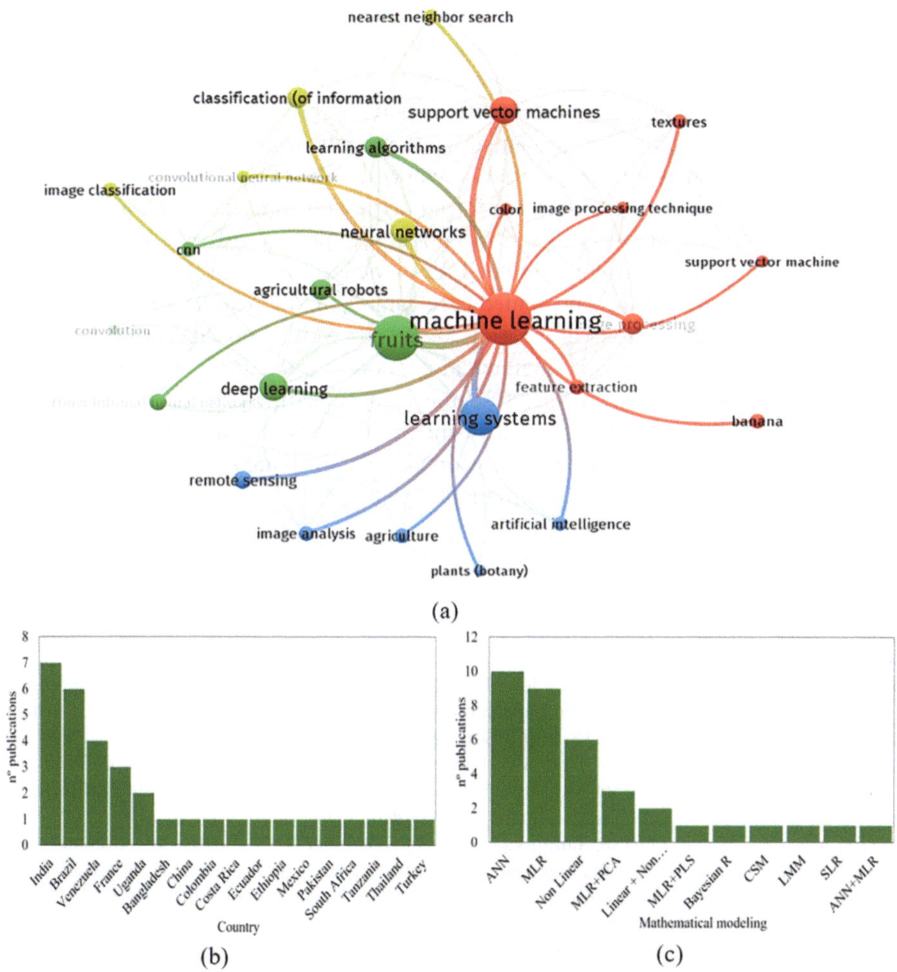

Fig. 1.9 (**a**) Co-occurrence of keywords with the VOSviewer software with clusters identified by different colors. (**b**) The number of publications by country. (**c**) The number of publications related to mathematical modeling on banana yield ($n = 36$). Note: artificial neural network (ANN), multiple linear regression (MLR), nonlinear regression, multiple linear regression and principal component analysis (MLR + PCA), linear and nonlinear regression, multiple linear regression and partial least squares (MLR + PLS), artificial neural networks and multiple linear regression (ANN + MLR), crop simulation models (CSM), Bayesian regression model, linear mixed models (LMM)

found. Four clusters are observed that represent four large thematic areas; the first cluster (red color in Fig. 1.9a) is words related to machine learning in bananas, and the second cluster (blue color in Fig. 1.9a) represents the learning system in agriculture, followed by the third cluster (green color in Fig. 1.9a) that is linked to deep learning in fruits and, finally, the cluster related to the application of classification algorithms (yellow color in Fig. 1.9a).

According to Jayasinghe et al. (2022) in 2020, the highest number of papers related to banana yield modeling attempts was reported, and India, Brazil, Venezuela, and France are the major countries where many scientists are interested in the topic (Fig. 1.9b). Multiple linear regression (MLR) approaches have been widely employed in the development of banana yield models, followed by machine learning techniques such as artificial neural networks (ANN) (Fig. 1.9c).

Mathematical models are decisive for strategic and forecast applications; however, banana-related models are less common, and reviews of previous modeling efforts are scarce, emphasizing the need for evidence-based studies on this important topic. This brief analysis presented here could help interested parties to identify the strengths and limitations of existing models as well as provide information on how to build novel and reliable mathematical models related to banana cultivation. In short, there is scope for further research using machine learning techniques in agronomy, which, within the context of this book, could contribute to identifying soil properties related to the incidence of diseases in bananas that could help to develop more sustainable banana production systems.

References

Ahmad F, Martawi NM, Poerba YS, de Jong H, Schouten H, Kema GH (2020) Genetic mapping of Fusarium wilt resistance in a wild banana *Musa acuminata* ssp. *malaccensis* accession. Theor Appl Genet 133(12):3409–3418. https://doi.org/10.1007/s00122-020-03677-y

Alimi OA, Ouahada K, Abu-Mahfouz AM, Rimer S, Alimi KOA (2021) A Review of Research Works on Supervised Learning Algorithms for SCADA Intrusion Detection and Classification. Sustainability 13(17):9597. https://doi.org/10.3390/su13179597

Aruraj A, Alex A, Subathra MSP, Sairamya NJ, George ST, Ewards SV (2019) Detection and classification of diseases of banana plant using local binary pattern and support vector machine. In: 2019 2nd international conference on signal processing and communication (ICSPC). IEEE, pp 231–235. https://doi.org/10.1109/ICSPC46172.2019.8976582

Bannerjee G, Sarkar U, Das S, Ghosh I (2018) Artificial intelligence in agriculture: A literature survey. Int J Comput Sci Appl 7(3):1–6

Baviera T (2017) Técnicas para el Análisis de Sentimiento en Twitter: Aprendizaje Automático Supervisado y SentiStrength. Revista Dígitos 1(3):33–50. https://doi.org/10.7203/rd.v1i3.74

Bouma J (2020) Contributing pedological expertise towards achieving the United Nations Sustainable Development Goals. Geoderma 375:114508. https://doi.org/10.1016/j.geoderma.2020.114508

Breiman L (2001) Random forests. Mach Learn 45(1):5–32. https://doi.org/10.1023/a:1010933404324

Breiman L, Friedman J, Stone CJ, Olshen RA (1984) Classification and regression trees. CRC Press

Calero J, Aranda V, Montejo-Raez A, Martin-Garcia JM (2018) A new soil quality index based on morpho-pedological indicators as a site-specific web service applied to olive groves in the Province of Jaen (South Spain). Comput Electron Agric 146:66–76. https://doi.org/10.1016/j.compag.2018.01.016

Coltro L, Karaski TU (2019) Environmental indicators of banana production in Brazil: *Cavendish* and *Prata* varieties. J Clean Prod 207:363–378. https://doi.org/10.1016/j.jclepro.2018.09.258

Companioni B, Mora N, Díaz L, Pérez A, Arzola M, Espinosa P, Hernández M, Ventura J, Pérez MC, Santos R, Lorenzo JC (2005) Identification of discriminant factors after treatment of resis-

tant and susceptible banana leaves with *Fusarium oxysporum* f. sp. *cubense* culture filtrates. Plant Breed 124(1):79–85. https://doi.org/10.1111/j.1439-0523.2004.00997.x
Cortes C, Vapnik V (1995) Support-vector networks. Mach Learn 20(3):273–297. https://doi.org/10.1007/BF00994018
Cortez A, Olivares B, Parra R, Lobo D, Rodríguez MF, Rey JC (2018) Descripción de los eventos de sequía meteorológica en localidades de la cordillera central, Venezuela. Cs Ing Apls I(1):22–44. https://doi.org/10.22206/cyap.2018.vlil.pp23-45
Delgado E, Gómez N, González O, Marín C (2008) Evaluación a nivel de finca del efecto de la alta densidad de siembra en plátano (*Musa* AAB cv. Subgrupo plátano Hartón), municipio Obispo, Barinas, Venezuela. Rev Fac Agron (LUZ) 25(4):27–38. https://n9.cl/poml2
Delgado E, Trejos J, Villalobos M, Martínez G, Lobo D, Rey J, Rodríguez G, Rosales F, Pocasangre L (2010a) Determinación de un índice de calidad y salud de suelos para plantaciones bananeras en Venezuela. Interciencia 35(12):927–933. https://n9.cl/nv7vv
Delgado E, Rosales F, Trejos J, Villalobos M, Pocasangre L (2010b) Índice de calidad y salud de suelos para plantaciones bananeras de cuatro países de América Latina y el Caribe. Bioagro 22(1):53–60. https://n9.cl/qm3e
Dita M, Perez-Vicente L, Guzmán M, Urias C, Staver C (2017) Fusarium tropical race 4 and its threat to food security and the banana industry in Latin America and the Caribbean. Phytopathology 107(7):8–8
Dita M, Barquero M, Heck D, Mizubuti ESG, Staver CP (2018) Fusarium wilt of banana: Current knowledge on epidemiology and research needs toward sustainable disease management. Front Plant Sci 9(1468):1–21. https://doi.org/10.3389/fpls.2018.01468
FAO (2019) Food outlook – biannual report on global food markets – November 2019. Rome https://www.fao.org/3/ca6911en/ca6911en.pdf
FAO (2021a) FAOSTAT Database: Area, production, and export of banana and plantain. http://wwwfaoorg/faostat/es/#data/QCL Accessed 21 Sept 2021
FAO (2021b) Perspectivas a mediano plazo: perspectivas para la producción y el comercio mundial de bananos y frutas tropicales 2019–2028. Roma. https://n9.cl/xfd23
FAO (2022) Banana Statistical Compendium 2021. Rome
Gauggel CA, Sierra F, Arévalo G (2003) The problems of banana root deterioration and their impact on production: production experience in Latin America. In: Turner DW, Rosales FE (eds) Banana root system: towards a better understanding for its productive management: Proceedings of an International Symposium/International Network for the improvement of banana and plantain. Montpellier. https://n9.cl/b3wi8
Ghag SB, Shekhawat UKS, Ganapathi TR (2015) Fusarium wilt of banana: biology, epidemiology, and management. Int J Pest Manag 61(3):250–263. https://doi.org/10.1080/09670874.2015.1043972
Gómez C, Surga J, Rumbos R, Rosales H, Vera J, Pino N, Castellano G, Marin C (2012) Efecto del postdeshoje para la reducción de *Mycosphaerella fijiensis*, en plátano cv. Hartón (*Musa* AAB), Venezuela. Rev Fac Agron (LUZ) 29(2):173–182. https://n9.cl/r5jqs
Gómez-Selvaraj M, Vergara A, Montenegro F, Alonso Ruiz H, Safari N, Raymaekers D, Ocimati W, Ntamwira J, Tits L, Omondi AB, Blomme G (2020) Detection of banana plants and their major diseases through aerial images and machine learning methods: A case study in DR Congo and the Republic of Benin. ISPRS 169:110–124. https://doi.org/10.1016/j.isprsjprs.2020.08.025
Hamed AA, Tawfik E, Abdel-Razik AB, Rashed MA, Abdel-Aziz SH, El-Shafie MA (2018) Production of Transgenic Cavendish Banana (*Musa Acuminata* L.) Resistant to Fungal Infection using Chitosan Nanoparticles. BJAS 3(1):131–137. https://doi.org/10.21608/BJAS2018.191091
Hearst MA, Dumais ST, Osuna E, Platt J, Scholkopf B (1998) Support vector machines. IEEE 13(4):18–28. https://doi.org/10.1109/5254.708428
Hernández Y, Portillo F, Portillo M, Navarro C, Rodríguez M, Velazco J (2006) Densidad estomática en materiales de plátano (*Musa* AAB, AAAB y ABB) susceptibles y resistentes a Sigatoka Negra (*Mycosphaerella fijiensis*, Morelet). Rev Fac Agron (LUZ) 23(3):294–300. https://n9.cl/gl9tq

Hou JC, Hu YH, Hou LX, Guo KQ, Satake T (2015) Classification of ripening stages of bananas based on support vector machine. Int J Agric Biol 8(6):99–103. https://doi.org/10.3965/j.ijabe.20150806.1275

Iskandar J, Kusmoro J, Mubarokah M, Partasasmita R (2018) Ethnobotany of banana plants (*Musa x paradisiaca*) of Palintang Hamlet, Cipanjalu Village, Bandung, West Java, Indonesia. Biodiversitas 19(6):2059–2072. https://doi.org/10.13057/biodiv/d190611

Jayasinghe S, Ranawana C, Liyanage I, Kaliyadasa P (2022) Growth and yield estimation of banana through mathematical modeling: A systematic review. J Agric Sci 160(3–4):152–167. https://doi.org/10.1017/S0021859622000259

Jha K, Doshi A, Patel P, Shah M (2019) A comprehensive review on automation in agriculture using artificial intelligence. Artif Intell Agric 2:1–2. https://doi.org/10.1016/j.aiia.2019.05.004

Jung J, Maeda M, Chang A, Bhandari M, Ashapure A, Landivar-Bowles J (2021) The potential of remote sensing and artificial intelligence as tools to improve the resilience of agriculture production systems. Curr Opin Biotechnol 70:15–22. https://doi.org/10.1016/j.copbio.2020.09.003

Kuhn M, Weston S, Culp M, Coulter N, Quinlan R (2018) Paquete C50. https://cran.r-project.org/web/packages/C50/C50.pdf

Lei Y (2017) Individual intelligent method-based fault diagnosis. In: Lei Y (ed) Intelligent Fault Diagnosis and Remaining Useful Life Prediction of Rotating Machinery (pp. 67–174). Butterworth-Heinemann. https://doi.org/10.1016/B978-0-12-811534-3.00003-2

Liaw A, Wiener M (2002) Classification and Regression by random Forest. R News 2:18–22. https://n9.cl/b57nr

Ma L, Fu T, Blaschke T, Li M, Tiede D, Zhou Z, Ma X, Chen D (2017) Evaluation of feature selection methods for object-based land cover mapping of unmanned aerial vehicle imagery using random forest and support vector machine classifiers. Int J Geo-Inf 6(2):51. https://doi.org/10.3390/ijgi6020051

Manjunath CB, Gnanappazham L, Vijayakumar RM, Kavino M, Jagadeeswaran R, Ramsubramoniam S (2019) Machine learning classifiers on sentinel-2 satellite image for the classification of banana (*Musa* Sp.) plantations of Theni district, Tamil Nadu, India. Int J Chem Stud 7(5):1419–1425. https://n9.cl/hdr2b

Martínez GE, Rey-Brina JC, Rodríguez D, Jiménez C, Rodríguez Y, Rumbos R, Pargas-Pichardo R, Martínez E (2020a) Análisis de la situación fitopatológica actual de las Musáceas comestibles en Venezuela. Agronomía Tropical 70:1–20. https://doi.org/10.5281/zenodo.4323273

Martínez G, Rey-Brina JC, Pargas-Pichardo RE, Domínguez M (2020b) Actualización sobre el manejo de la marchitez por Fusarium Raza 4 Tropical en musáceas. Agronomía Tropical 70:1–18. https://doi.org/10.5281/zenodo.4291108

Menéses A, Pocasangre LE, Somarriba E, Riveros AS, Rosales FE (2003) Diversidad de hongos endofíticos y abundancia de nematodos en plantaciones de banano y plátano de la parte baja de los territorios indígenas de Talamanca. Agroforestería de las Américas 10(37):59–62. https://n9.cl/gqy7y

Nava C (1997) El plátano, su cultivo en Venezuela. Ediciones Astro Data S.A, Maracaibo, Venezuela, 134 p

Olivares B, Cortez A, Parra R, Lobo D, Rodríguez MF, Rey JC (2017) Evaluation of agricultural vulnerability to drought weather in different locations of Venezuela. Rev Fac Agron (LUZ) 34(1):103–129. https://n9.cl/d827w

Olivares B, Paredes F, Rey J, Lobo D, Galvis-Causil S (2021a) The relationship between the normalized difference vegetation index, rainfall, and potential evapotranspiration in a banana plantation of Venezuela. SAINS TANAH 18(1):58–64. https://doi.org/10.20961/stjssa.v18i1.50379

Olivares BO, Calderón AR, Rey JC (2021b) Clasificación de zonas afectadas por la marchitez en banano: una aplicación con algoritmos de Machine Learning en Venezuela. REICIT 1(1):1–7

Olivares B, Vega A, Calderón MAR, Rey JC, Lobo D, Gómez JA, Landa BB (2022a) Identification of Soil Properties Associated with the Incidence of Banana Wilt Using Supervised Methods. Plan Theory 11(15):2070. https://doi.org/10.3390/plants11152070

Olivares BO, Vega A, Rueda Calderón MA, Montenegro-Gracia E, Araya-Almán M, Marys E (2022b) Prediction of Banana Production Using Epidemiological Parameters of Black Sigatoka: An Application with Random Forest. Sustainability 14:14123. https://doi.org/10.3390/su142114123

Olivares BO, Rey JC, Perichi G, Lobo D (2022c) Relationship of Microbial Activity with Soil Properties in Banana Plantations in Venezuela. Sustainability 14:13531. https://doi.org/10.3390/su142013531

Owomugisha G, Quinn JA, Mwebaze E, Lwasa J (2014) Automated vision-based diagnosis of banana bacterial wilt disease and black Sigatoka disease. In: International conference on the use of mobile ICT in Africa (pp. 1–5). https://n9.cl/ftc3n

Pattison T, Badcock K, Lindsay S, Armour A, Velupillai R, Moody P, Smith L, Gullino L, Cobon J (2004) Banana root and soil health project – field workbook. Department of Primary Industries and Fisheries, Queensland, 15 p. https://n9.cl/nrtvc

Pattison AB, Wright CL, Kukulies TL, Molina AB (2014) Ground cover management alters development of Fusarium wilt symptoms in Ducasse bananas. Australas Plant Pathol 43(4):465–476. https://doi.org/10.1007/s13313-014-0296-5

Ploetz RC (2015) Management of Fusarium wilt of banana: A review with special reference to tropical race 4. Crop Prot 73:7–15. https://doi.org/10.1016/j.cropro.2015.01.007

Pocasangre LE, Sikora RA, Vilich V, Schuster RP (2000) Survey of banana endophytic fungi from Central America and screening for biological control of *Radopholus similis*. Acta Hortic 531:283–289. https://doi.org/10.17660/ActaHortic.2000.531.47

Quinlan JR (2007) Decision trees as probabilistic classifiers. In: Proceedings of the fourth international workshop on machine learning. Morgan Kaufmann Publishers, Inc, Massachusetts. https://doi.org/10.1016/b978-0-934613-41-5.50007-6

Quinlan JR (2014) C4. 5: programs for machine learning. Elsevier, Amsterdam

Rey JC, Chacín M, Sapuky M, Núñez M, Martínez G, Rodríguez G, Espinoza J, Arturo M, Pocasangre L, Delgado E, Rosales F (2006) Aptitud de las tierras para banano en suelos de Venezuela y su relación con la productividad. XVII Reunión Internacional ACORBAT: Banano un negocio sustentable. Joinville. Santa Catarina, Brasil. Nov 15–20. p. 362

Rey JC, Martínez G, Lobo D, Trejos J, Pocasangre L, Rosales F (2009) Aspectos sobre calidad y salud de suelos bananeros en Venezuela. Producción Agropecuaria 2:52–55. https://n9.cl/zthj2

Rey JC, Martínez-Solórzano G, Ramírez H, Pargas-Pichardo R (2020) Relación de las condiciones agroecológicas de un lote de planicie lacustrina con la marchitez del banano Cavendish en Aragua, Venezuela. Agronomía Tropical 70:1–12. https://doi.org/10.5281/zenodo.4346252

Rosales FE, Pocasangre LE, Trejos J, Serrano E, Peña W (2008) Guía de Diagnóstico de la Calidad y Salud de Suelos. Bioversity International, Montpellier, Francia, 79 pp

Sabiiti G, Ininda JM, Ogallo LA, Ouma J, Artan G, Basalirwa C, Nanteza J (2018) Adapting agriculture to climate change: suitability of banana crop production to future climate change over Uganda. In: Limits to climate change adaptation. Springer, Cham, pp 175–190. https://doi.org/10.1007/978-3-319-64599-5_10

Sangeetha T, Lavanya G, Jeyabharathi D, Rajesh Kumar T, Mythili K (2020) Detection of pest and disease in the banana leaf using convolution random forest. Test Eng Manag 83(2):3727–3735. https://n9.cl/yeffq

Scott GJ (2021) A review of the root, tuber, and banana crops in developing countries: Past, present, and future. Int J Food Sci 56(3):1093–1114. https://doi.org/10.1111/ijfs.14778

Serrano E (2003) Relación entre los contenidos de raíz funcional y la productividad de banano en Costa Rica. In: Turner DW, Rosales FE (eds) 2005. Banana Root System: towards a better understanding for its productive management: Proceedings of an international symposium. International Network for the Improvement of Banana and Plantain, Montpellier, p 28

Van Eck NJ, Waltman L (2010) Software survey: VOSviewer, a computer program for bibliometric mapping. Scientometrics 84(2):523–538. https://doi.org/10.1007/s11192-009-0146-3

Van Engelen JE, Hoos HH (2020) A survey on semi-supervised learning. Mach Learn 109(2):373–440. https://doi.org/10.1007/s10994-019-05855-6

Vapnik VN (1995) The nature of statistical learning theory. Springer-Verlag, Berlin, Heidelberg
Varma V, Bebber DP (2019) Climate change impacts on banana yields around the world. Nat Clim Chang 9(10):752–757. https://doi.org/10.1038/s41558-019-0559-9
Wang X, Lin X, Dang X (2020) Supervised learning in spiking neural networks: A review of algorithms and evaluations. Neural Netw 125:258–280. https://doi.org/10.1016/j.neunet.2020.02.011
Xanthopoulos P, Pardalos PM, Trafalis TB (2013) Linear discriminant analysis. In: Robust data mining. Springer, New York, NY, pp 27–33. https://doi.org/10.1007/978-1-4419-9878-1_4
Ye HC, Huang WJ, Huang SY, Cui B, Dong YY, Guo AT, Ren Y, Jin Y (2020) Identification of banana fusarium wilt using supervised classification algorithms with UAV-based multi-spectral imagery. Int J Agric Biol Eng 13(3):136–142. https://doi.org/10.25165/j.ijabe.20201303.5524

Chapter 2
Methodological Framework

Abstract The methodological framework adopted for this study in general, the specific methodology used for the four scientific studies (Chaps. 3–6) related to plant health and banana productivity in Venezuela, as well as the research strategies that led to the results are developed searched results. This research was structured according to four ways of approaching the study in banana areas. This chapter describes the application of different machine learning algorithms in banana production systems in Venezuela through four different topics: (a) a systematic review and bibliometric analysis of the agro-environmental factors that determine the appearance of *Fusarium oxysporum* f. sp. *cubense* Tropical Race 4 (*Foc* TR4) and the risk analysis and climatic suitability maps for (syn. *Fusarium odoratissimum*) in Venezuela through the maximum entropy model; (b) the relationship between soil properties and the incidence of banana wilt (MB) through Orthogonal Partial Least Squares Discriminant Analysis (OPLS-DA) and the random forest (RF) algorithm; (c) an analysis of the relationship between main physical, chemical, and biological soil properties most correlated to banana productivity; and (d) the potential use of soil morphological properties to differentiate banana productivity levels under Venezuelan conditions through a prediction model based on categorical soil properties with regularized optimal scaling regression.

Keywords Methodology · Research strategies · Technological organization · Conceptualization · Formal analysis · Data curation

2.1 Introduction

One of the great advances of machine learning in agriculture is the analysis of the productive activity of agri-food marketers to adjust the efficiency and precision indices in their operations (Larada et al. 2018; Manrique-Silupu et al. 2021). However, it is not the only step in the food chain where this technology performs with prowess (Singh et al. 2022).

B. O. Olivares Campos, *Banana Production in Venezuela*, The Latin American Studies Book Series, https://doi.org/10.1007/978-3-031-34475-6_2

In the banana sector, the fields in which it is applied include the optimization of production and commercialization, the projections of annual productive results, and the anticipation of errors based on data that are obtained from each farm cultivated, that is, with particular analysis (Pitti et al. 2021; Montenegro et al. 2021). The advantages of the predictions achieved, thanks to the "learning" machines, include improving the value of future crops, reducing costs in R&D techniques (research and development), knowing the climatic suitability, modifying production decisions based on soil data or disease incidence, and, finally, optimizing production processes (Olivares et al. 2020a, b). In addition to making decisions to avoid bottlenecks in production and being able to generate a detailed mapping of cultivated areas, machine learning in agriculture offers an advance that guarantees decision-making intending to achieve sustainability.

Experts point to a reduction of up to 90% of phytosanitary products, the strengthening of plant health in a natural way, and a 20% increase in crop yields (Espejo et al. 2018; de Oliveira Aparecido et al. 2020). Predictive learning from algorithms drastically reduces resource consumption, which not only means savings for the farmer but also the possibility of creating responsible and effective consumption awareness.

As mentioned in the Introduction, this research was structured according to two themes to address novel solutions to banana productivity and plant health in Venezuela. On the one hand, the study of Fusarium wilt of bananas was addressed based on a literature review, a bibliometric analysis, and a climatic suitability model for the country, followed by the application of machine learning algorithms to identify soil properties associated with banana wilt. On the other hand, to recognize numerical and categorical properties of the soil related to banana productivity, the research included studies with a statistical approach to show the usefulness of this type of analysis.

2.2 Study of Fusarium Wilt of Bananas in Venezuela

Historically, *Fusarium oxysporum* f. sp. *cubense* (*Foc*), the causal agent of Fusarium wilt (FW) of bananas, is the main threat for different banana cultivars of economic and strategic importance worldwide. Detailed analysis and description of FW of bananas or its causal agent (*Foc*) have been thoroughly reviewed in the past (Stover 1962; Nelson 1991; Ploetz 1994; Bentley et al. 1998; Ploetz and Pegg 2000), and more recently (Davis 2005; Pérez-Vicente 2004; Ploetz 2006, 2015a,b; Pocasangre et al. 2011; Ghag et al. 2015; Dita et al. 2018; Bubici et al. 2019; Kema et al. 2021), although most of the analysis has devoted limited attention to the interaction between FW and agro-environmental factors. Tropical Race 4 (TR4) is the most virulent race of *Foc* (currently assigned to a new species, *Fusarium odoratissimum*).

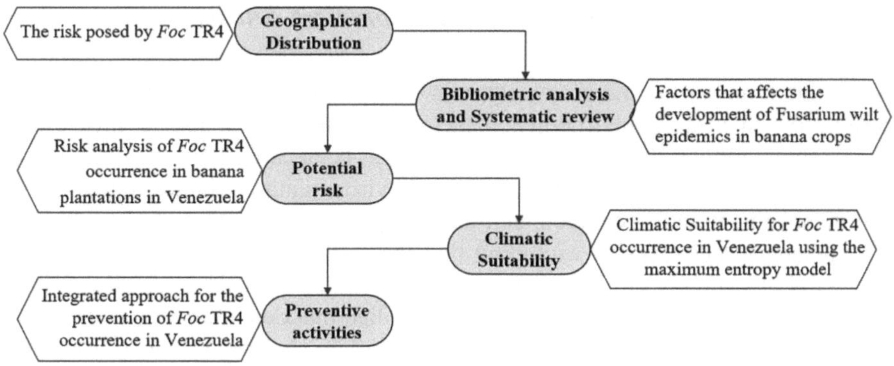

Fig. 2.1 General scheme of the study of Fusarium wilt of bananas in Venezuela

Foc is a pathogen that inhabits the soil and produces three types of asexual spores: microconidia, macroconidia, and chlamydospores (Nelson 1991). The chlamydospores are highly resistant double membrane propagules that allow the pathogen to remain viable in the soil for many years in the absence of a host, making it not possible to replant susceptible cultivars in the same soil once infested (Davis 2005).

Three races of *Foc* (Race 1 (R1), Race 2 (R2), and Race 4 (R4)) and several vegetative compatibility groups (VCG) within each race differing in virulence have been described in the populations of this pathogen (Dita et al. 2018; Bentley et al. 1998; Ploetz 2006). *Foc* R1 is responsible for the epidemic in "Gros Michel" and "Manzano" clones, in addition to "Pome" (AAB), "Pisang Awak" (ABB), and "Maqueño" (AAB). *Foc* R2 especially attacks bananas belonging to the subgroup "Bluggoe" or "Topocho" (ABB) (Ploetz 2006). *Foc* R4 is the most dangerous of all races because it attacks all these groups of banana plants, including the Cavendish clone (Stover 1962; Ploetz 2015a).

In this study, according to the approach (Fig. 2.1), first, a geographical distribution was made; then, agri-environmental factors are described, specifically the factors that affect the development of FW epidemics in banana crops; after, the risk analysis of *Foc* TR4 occurrence in banana plantations in Venezuela was made; then, climatic suitability for *Foc* TR4 occurrence in Venezuela using the maximum entropy model was obtained; and finally, the activities aimed at preventing the entry and spread of *Foc* TR4 in the banana production systems of Venezuela are described.

2.2.1 *Geographical Distribution of Foc*

A brief description of the global distribution of *Foc* races is provided according to CABI/EPPO (2015), EPPO (2021), and Promusa (2022) registries, emphasizing the risk of entry into Venezuelan banana areas.

2.2.2 A Systematic Review and Bibliometric Analysis

An updated, reflective, and practical review of the current state of knowledge concerning the main agro-environmental factors that may affect disease progression and dissemination of this dangerous pathogen has been carried out, focusing on the Venezuelan Musaceae production systems. Environmental variables together with soil management and sustainable cultural practices are important factors affecting FW incidence and severity, excluding that the widespread dissemination of *Foc*, especially of its highly virulent tropical race 4, is mainly caused by human activities.

In this study, the statistical method was applied that quantitatively analyzes the research works related to *Foc* in Latin America and the caribbean (LAC) through the mathematical method known as bibliometrics (Van Eck and Waltman 2010), which can provide an overview of the quality of the studies, analyze the key areas of research, and predict the direction of future studies. To date, no bibliometric analysis of publications on *Foc* in LAC has been published. As this lethal disease in bananas has not been completely under control and more knowledge must be obtained from these references, a bibliometric analysis of it is a critical need. Therefore, our study was done in time to provide a broad understanding of *Foc* in LAC and future research directions.

Literature on *Foc* TR4 published between 2019 and 2022 related to LAC was extracted from the Web of Science (WOS) collection database. Keywords such as "*Fusarium oxysporum* f. sp. *cubense*," "Fusarium wilt Musa," and "Foc TR4" were used to reach the relevant publications, delimiting the studies of Latin American and Caribbean countries. VOSviewer was applied to perform the bibliometric analysis of these articles (Van Eck and Waltman 2010).

2.2.3 Potential Risk Analysis and Climatic Suitability for Foc TR4

Additionally, risk analysis and climatic suitability maps for *Foc* TR4 in Venezuela have been developed. Maximum entropy (MaxEnt) models of machine learning are used to model the potential distribution of organisms to obtain predicted new places of occurrence. It identifies areas that have conditions most like species' current known occurrences and ranks them from "0" (unsuitable or most dissimilar) to "1" (most suitable or most similar). In this perspective, the MaxEnt model v. 3.4.3 (Phillips et al. 2006) was used to estimate the potential for the establishment of Foc TR4 in Venezuela. Presence-only data was obtained from the current known global distribution of *Foc* TR4 obtained from the following sources: (i) the CABI Crop Protection Compendium (CABI/EPPO 2015), (ii) the ProMusa project website (Promusa 2022), and (iii) from reviewing the specialized literature (Olivares et al. 2021b). We only selected references that contained geographical information about the presence of the pathogen.

Climate data were obtained from Chelsa Climatology (Karger et al. 2017) that includes monthly mean temperature and precipitation patterns for the 1979–2013 period. 19 bioclimatic variables were derived from monthly temperature and precipitation values *sdm* package in *R* software version 4.0.2 (R Core Team 2020) using a fivefold cross-validation procedure.

2.2.4 Activities Aimed at Preventing the Entry and Spread of Foc TR4

Both the strategies and the plan of priority activities on Foc TR4 were built with the collaboration of the FAO and experts and technicians from several countries in the Latin American region. These activities were aligned with the FAO Strategic Framework for 2022–2030 (FAO 2021); the priorities of the Sendai Framework and the Regional Strategy for Disaster Risk Management in the Agricultural Sector and Food and Nutritional Security in Latin America and the Caribbean (UNISDR 2015); consider the International Standards for Phytosanitary Measures (IPPC 2021) and introduce innovative elements for the management of phytosanitary emergencies in Venezuela.

2.3 Study of Banana Wilt Incidence in Venezuela

Banana wilt (BW) is known as False Panama Disease; it was first described by Deacon et al. (1985), which can easily be confused with Fusarium wilt. Recently, BW is associated with a fungal-bacterial complex. The literature and different studies developed in Venezuela (Martinez et al. 2020; Rey et al. 2020; Olivares et al. 2021a) suggest that this disease is a situation of plant stress caused by abiotic factors, which would enhance the action of certain microorganisms that could have a determining role in the expression of symptoms. In this study, the Orthogonal Partial Least Squares Discriminant Analysis (OPLS-DA) and the random forest (RF) algorithms were applied to find soil variables capable of distinguishing banana lots showing high or low BW incidence.

2.3.1 Study Area

The study was carried out in a banana plantation located in the Aragua state, with 205 ha planted with Cavendish cv. Pineo Gigante (67.58° W, 10.14° N; Fig. 2.2). These plants had at the time of sampling (i) a leaf number from 16 to 18, (ii) height values ranging from 3.5 to 4.5 m, and (iii) a growth period from 9 to 10 months.

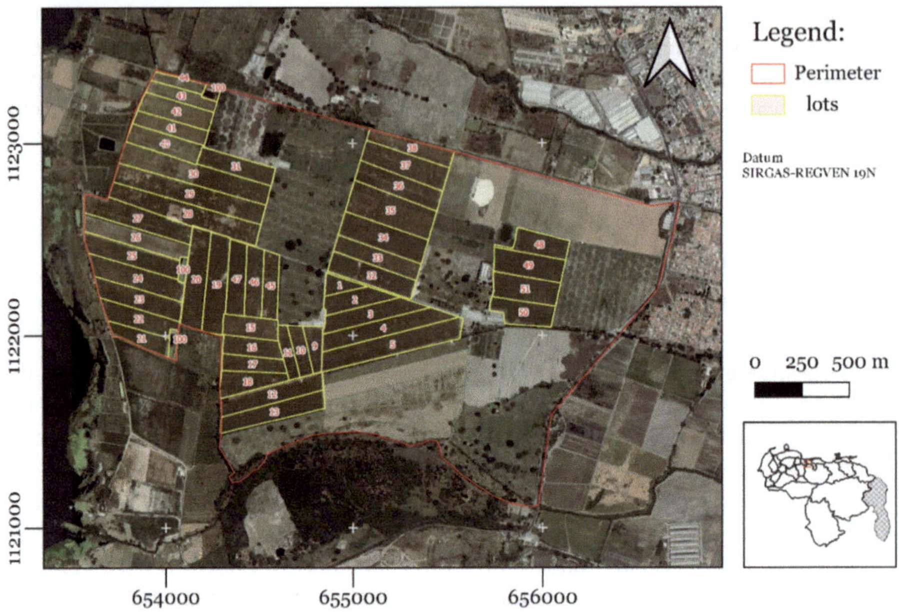

Fig. 2.2 Geographical location of the study area with banana lots (Olivares et al. 2022)

This region is characterized by a Tropical Savanna climate (Aw). The annual mean rainfall is 980 mm (Olivares et al. 2020a, b) and shows a marked seasonal pattern with a wet season from May to October. The mean annual temperature is 26.2 °C, whereas the mean annual relative humidity is 70.0% (Olivares 2018). The terrain relief is mostly flat (slope ranging from 0% to 2%). The predominant types of soil are Mollisol and Entisol, which are mostly of lacustrine origin, with medium textures, high nutrient availability, moderate to good drainage, soil pH varying from neutral to alkaline, good fertility, and high soil organic matter content (Delgado et al. 2010).

2.3.2 *Soil Sampling*

A systematic soil sampling was carried out in 39 banana lots sampled in January 2016 and 2017 (total banana lots sampled, $n = 78$). Sampling was conducted following the guidelines of Lozano et al. (2004), with an approximate distance of 150 m between sampling sites. Composite soil samples were obtained from each of the banana lots, in the first horizon at a depth of 0 to 20.0 ± 5.0 cm. The samples were subjected to soil analysis for fertility characterization purposes; in total, 16 soil variables were determined including the percentage of sand, silt, and clay (Gee and

Or 2002); soil reaction (pH); electrical conductivity (EC, dS/m) in 1:2 (soil/water) suspension (Soil Survey Staff 2017); organic matter (OM, %) (Heanes 1984); and available contents of potassium (K, mg/kg), sodium (Na, mg/kg), magnesium (Mg, mg/kg), calcium (Ca, mg/kg), manganese (Mn, mg/kg), iron (Fe, mg/kg), zinc (Zn, mg/kg), copper (Cu, mg/kg), sulfur (S, mg/kg), and phosphorus (P, mg/kg) (Mehlich 1984).

2.3.3 Banana Wilt Incidence

Before the beginning of the study, plants with typical symptoms of BW were located and identified in all the lots of the farm, from which tissue samples were taken from the pseudostem and roots for the identification of pathogenic microorganisms. The isolation method in a Potato Dextrose Agar (PDA) culture medium and the humid chamber was used in the laboratory of the Faculty of Agronomy of the Central University of Venezuela.

For the identification of BW incidence in the field, in each banana lot, each banana plant was individually inspected monthly for the presence of symptoms compatible with BW. Banana plants showing BW symptoms were eliminated in each lot and each evaluation period. Therefore, in the next monthly inspection, only the number of plants with new BW symptoms to that date was counted. The cumulative incidence of BW was determined in each of the 78 banana lots sampled in 2016 and 2017 using the guidelines by Bosman (2016). The main aim of the continuous monitoring of BW incidence was to determine the new cases of BW that occur in the total population of plants planted in each banana lot and sampling time (Eq. 2.1). Hence, the cumulative incidence rate is calculated as the sum of the monthly incidence of BW values in percentage for each banana lot in a particular year.

$$\text{Cumulative incidence rate}(\%) = 100 \times \sum_{i=1}^{12} \frac{n^\circ\text{of diseased plants}(BW)}{\text{Total plants planted}} \tag{2.1}$$

In Venezuela there is no information describing threshold values to establish categories for BW incidence. Therefore, we established a threshold value of 1.90% for BW incidence that allowed classifying the lots as having low (<1.90% of affected plants) or high (≥1.90% of affected plants) incidence of BW in these lacustrine soils. This high incidence value would represent a decrease of up to 13,300 kg/ha/year in those bananas lots showing an incidence of BW of 1.90% and was selected based on the information provided by J.C. Rey (personal communication, September 28, 2019) and several years of experience observing yield losses associated to BW.

2.3.4 Data Analysis

2.3.4.1 Exploratory Analysis

The objective of this exploratory analysis was to select the algorithm with the best performance in the classification of areas affected by BW by using machine learning algorithms such as random forest (RF) (Eq. 1.1 in Chap. 1), analysis linear discriminant function (ADL) (Eq. 1.2 in Chap. 1), radial basis function kernel Support Vector Machines (RSVM) (Eq. 1.3 in Chap. 1), linear kernel Support Vector Machines (LSVM) (Eq. 1.4 in Chap. 1), classification trees (CART), and the decision tree algorithm (C5.0) (Eq. 1.5 in Chap. 1).

To address the issue of imbalance, present in the BW incidence variable (different amounts of banana lots with high and low incidence), data pre-processing techniques known as resampling were applied. These techniques consist of making direct modifications to the elements of the database to generate a balance between the different classes that compose it. Four variants of resampling techniques were used in this exploratory analysis: undersampling (Guo et al. 2017), oversampling (Estabrooks et al. 2004; Liu et al. 2007), random oversampling examples (ROSE) (Lunardon et al. 2014), and synthetic minority oversampling technique (SMOTE) (Chawla et al. 2002). The undersampling technique works on the majority class of the dataset and consists of eliminating samples within this class in such a way as to generate a balance between classes (Guo et al. 2017). On the other hand, the oversampling technique, contrary to the one previously named, performs its work on the minority class of the dataset, adding samples to this class in such a way that the difference between classes is balanced (Estabrooks et al. 2004; Liu et al. 2007). ROSE produces a synthetic, possibly balanced, a sample of data simulated according to the Bootstrap smoothed resampling approach (Lunardon et al. 2014). Finally, SMOTE is a technique based on nearest neighbors judged by the Euclidean distance between data points in the feature space (Chawla et al. 2002).

The train function of the R library "Caret" was used to fit the algorithms under different tuning parameters, and the best model was selected using as criterion the highest value of the area under the receiver operating characteristic (ROC) curve. Classification of banana lots comprises two major steps: training and testing processes. In this work, the k-fold cross-validation method is used to obtain the training and the testing datasets (Salzberg 1997). To analyze the performance of the proposed algorithms, RF, LDA, CART, RSVM, LSVM, and C5.0 classifiers with five-fold cross validation are used in this work. The estimators used as criteria to measure the performance of the algorithms were accuracy, sensitivity, specificity, and Cohen's kappa (Feuerman and Miller 2008).

2.3.4.2 Formal Data Analysis

After having carried out the exploratory analysis for the selection of the algorithm with the best performance with the resampling techniques, the formal data analysis was carried out. Before formal data analysis, we checked the data integrity. The normalization of the soil variables was carried out using the statistical package in R software version 4.0.2 (R Core Team 2020) based on the geometric mean, and a generalized logarithmic transformation using the "glog" function in R was performed to make the variables comparable among them due to differences on units to measure them (Chong et al. 2019; Yang et al. 2021). Figure 2.3 shows the general scheme of the data analysis procedures followed in this work.

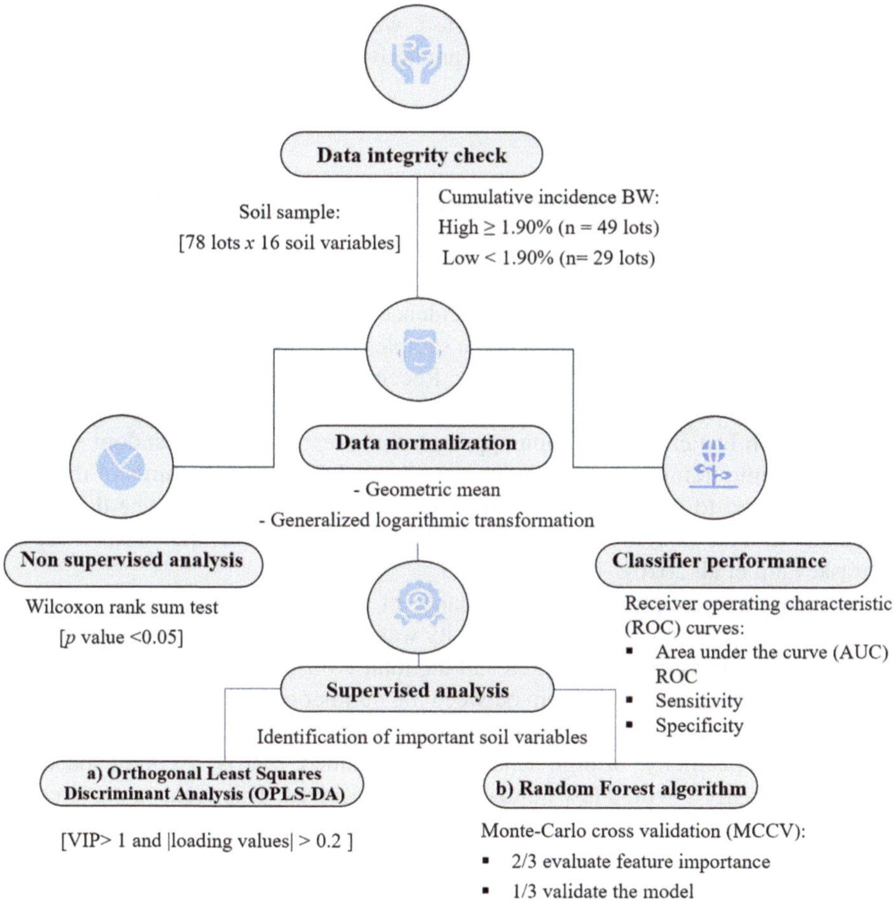

Fig. 2.3 General scheme of the data analysis procedures (sample size, $n = 78$; variable size, $n = 16$)

2.3.5 *Identification of Important Soil Variables*

For identification of relevant soil variables characterizing the incidence of BW, a Wilcoxon rank sum test was performed to find the most important features of soil variables at a threshold *p* value <0.05 (Chong et al. 2019) showing differences between the groups of banana lots with low and high incidence. Next, an Orthogonal Partial Least Squares Discriminant Analysis (OPLS-DA) was used to reduce the number of soil variables in high-dimensional data to produce a robust and easy-to-interpret model and to identify the main soil characteristics that drive the separation of plant lots based on BW incidence (Low or high). This multivariate statistical analysis was carried out using "ropls" R packages (Bylesjö et al. 2006).

The variable importance in projection (VIP) >1 and corresponding |loading values| >0.2 in the model were used to identify the variables responsible for distinguishing both BW categories (Szymańska et al. 2012). Furthermore, a permutation test with 100 permutations was employed to validate the performance of the OPLS-DA model. For the quality criterion, we chose the OPLS-DA model, R^2Y (goodness-of-fit parameter), and Q^2 (predictive ability parameter) >0.5 (Triba et al. 2015).

2.3.6 *Classifier Performance and Accuracy Assessment*

The random forest (RF) algorithm was used as a machine learning approach for classifying lots with a high and low incidence of BW (Breiman et al. 2003). RF models allow the prediction of unknown samples (i.e., test dataset) after training on a known dataset (i.e., training dataset). Receiver operating characteristic (ROC) curves were generated by Monte Carlo cross validation (MCCV) (Xu and Liang 2001), which is a cross-validation approach that creates multiple random splits of the dataset into training and validation data. In each MCCV, two-thirds of the samples were used to evaluate feature importance, and the remaining one-third were used to validate the model created in the first step (Xia and Wishart 2016; Paraskevaidi et al. 2020).

To determine the predictive performance of the model, the graphs of the ROC curve were used, from which the sensitivity was defined as the relationship between the number of P correctly classified and the total P observed, against "1 – specificity" (specificity is the relationship between the number of N correctly classified and the total N observed). A model will have high predictive performance if at low values of "1 – specificity," a high sensitivity is obtained, that is, a good capacity to correctly classify P with a low number of false positives. This yields a curve closer to the upper left corner (Garosi et al. 2019). The area under the ROC curve (AUC) quantifies this relationship so that a model is considered acceptable if AUC ≥ 0.7, excellent if AUC ≥ 0.8, and outstanding if AUC ≥ 0.9.

2.4 Study of Main Physical, Chemical, and Biological Soil Properties with the Banana Productivity

The quality and health of soils can be measured using indicators, which are nothing more than soil variables that allow evaluation of their condition and carry information about changes or trends in that condition (Dumanski et al. 1998). Soil quality indicators can be physical, chemical, and biological properties, or processes that occur in them (SQI 1996). Therefore, it is widely accepted that the measurement of soil quality can be established from appropriate indicators, which reflect essential processes (physical, chemical, and biological) that take place in the soil, while at the same time being sensitive to detect differences in space and time, clearly establishing a cause-effect relationship.

In this study, the random forest algorithm was designed and applied to select the most important variables of a mathematical model that describes the quality and health of banana soils in Venezuela, indicating the most critical factors in the production of the studied sites. The collected data describe the chemical, physical, and microbiological components coming from the high- and low-productivity sectors.

2.4.1 Study Areas

Six banana fields located in the states of Aragua and Trujillo in Venezuela were selected (Table 2.1). The banana plantations of the Aragua state (fields A1, A2, A3, and A4) are in the Valencia Lake Basin, characterized by a Tropical Savanna (Aw) type of climate with an average annual rainfall depth of 980 mm (Olivares et al. 2018). Rains in this area are seasonal, concentrated between May and October. The average mean annual temperature value is 26.2 °C with an average annual relative humidity of 70.0% (Olivares 2018). The terrain relief is flat (slope ranging from 0% to 2%). The predominant soil types are Mollisol and Entisol, with moderate to good drainage, soil pH varying from neutral to alkaline, good fertility, and medium to high organic matter (Delgado et al. 2010).

The second study area is located at the Trujillo state (fields T1 and T2), in the Southeast Region of Lake Maracaibo, and is also characterized by a Tropical dry-winter (Aw) type of climate. The average annual precipitation is 950 mm, with two marked wet periods (dry season and rainy season) (Olivares et al. 2017a, b). Its annual mean temperature is 27.5 °C, and the annual average relative humidity is 78.0%. This area of the Trujillo state is an alluvial plain with slopes below 1% and mainly Entisol soils. The drainage is moderate to poor with neutral to alkaline pH soils, with moderate fertility and average organic matter content, around 2.75% (Rodríguez et al. 2006; Rey et al. 2009).

For field sampling, areas of high and low productivity were identified in each study area based on farmers' information and preliminary productivity assessment combined with data on the potential production and vigor obtained from each

Table 2.1 Geomorphological description and taxonomic classification of banana soils sampled in the study in each state by productivity level (Olivares et al. 2020a, b)

Fields	Productivity level[a]	Geographic coordinates	Geomorphological environment				Soil taxonomy[b]
			Natural region	State	Relief	Slope (%)	
Aragua state							
A1	HH	10°08′52″N 67°35′23″W	Depression of Valencia Lake	Aragua	Lacustrine plain	0.5–1.0	Mollic Ustifluvents
	HL	10°08′55″N 67°35′13″W				0.5–1.0	Fluventic Haplustolls
	LL	10°08′44″N 67°35′09″W				0.5–1.0	Fluventic Haplustolls
	LH	10°08′52″N 67°35′03″W				1.0–2.0	Aquic Ustifluvents
A2	HH	10°12′36″N 67°24′44″W	Depression of Valencia Lake	Aragua	Terrace	0.5–1.0	Fluventic Haplustepts
	LL	10°12′55″N 67°23′43″W				2.0	Fluventic Haplustolls
A3	HH	10°11′30″N 67°31′06″W	Depression of Valencia Lake	Aragua	Alluvial plain	0.5	Fluventic Haplustolls
	LL	10°11′38″N 67°31′06″W				0.5	Fluventic Haplustolls
A4	HH	10°11′24″N 67°31′26″W	Depression of Valencia Lake	Aragua	Alluvial plain	0.5	Fluventic Haplustolls
	LL	10°11′19″N 67°31′27″W				0.5–1.0	Fluventic Haplustolls
Trujillo state							
T1	HH	09°29′14″N 70°57′05″W	Recent alluvial depression of Lake Maracaibo	Trujillo	Alluvial plain	0.5–1.0	Oxyaquic Ustifluvents
	HL	09°29′55″N 70°56′50″W				0.5–1.0	Typic Ustifluvents
	LL	09°30′20″N 70°57′12″W				0.5	Typic Ustifluvents
	LH	09°30′24″N 70°57′32″W				1.0	Fluvaquentic Haplustolls
T2	HH	09°29′09″N 70°56′25″W	Recent alluvial depression of Lake Maracaibo	Trujillo	Alluvial plain	0.5–1.0	Oxyaquic Ustifluvents
	HL	09°28′32″N 70°55′53″W				0.5–1.0	Fluaquentic Haplustolls
	LL	09°29′08″N 70°56′14″W				0.5–1.0	Typic Ustifluvents
	LH	09°29′02″N 70°56′09″W				0.5–1.0	Oxyaquic Ustifluvents

[a]Classification nomenclature of banana soil productivity according to average fruit production: HH = High-High yield (≥40 t/ha/year); HL = High-Low yield (35–40 t/ha/year); LL = Low-Low yield (≤30 t/ha/year); LH = Low-High yield (30–35 t/ha/year)

[b] Soil Survey Staff (2010)

banana site from available records. The final location and number of sampling sites were determined according to the size of the farm.

To estimate banana productivity, sampling was performed according to the guidelines proposed by Rosales et al. (2008). In large fields (≥50 ha, fields A1, T1, and T2), four yield levels of "Grand Nain" bananas, determined previously based on the productivity history of all Venezuelan farms, were identified. These levels were High-High (HH) (≥40 t/ha/year), High-Low (HL) (35–40 t/ha/year), Low-Low (LL) (≤30 t/ha/year), and Low-High (LH) (30–35 t/ha/year). In these large field areas, a plot for each of these four yield levels was identified and established according to average fruit production. Each plot has an area of 4 ha with four repetitions sampled in each plot. In the remaining small fields (<25 ha, fields A2, A3, and A4), only two plots, with a surface area of 1 ha each, were identified: High-High (≥30 t/ha/year) and Low-Low productivity (<30 t/ha/year), with two repetitions sampled per plot. In this way, the indicators were measured and evaluated for their discriminatory quality in two productive environments at the macro level (differences among fields) and the micro level (differences within the fields).

2.4.2 *Production System Description*

2.4.2.1 Field A1

Field A1 has a cultivated area of 135 ha; 100% of the production is aimed at the local market. However, the technological reference is like that of an export farm. Fertilization is done by applying chemical fertilizers (27 cycles/year of complete formulas and 16 cycles/year of foliar fertilizers) and manure compost as organic fertilizer. In addition, calcium sulfate is applied as a salinity correction.

Phytosanitary management is carried out by applying systemic fungicides (9 cycles/year) and protectant fungicides (9 cycles/year). For weed control, post-emergent herbicides (e.g., Gramoxone, Finale, and Glyphosate) are applied in 12 cycles/year; pre-emergent herbicides are also used. For the control of nematodes, organophosphate nematicides are applied in some banana lots. To control insects, insecticides are used, applied in covers and biological repellents, as well as traps to control weevils.

As cultural practices for disease management (defoliation, topping, weed control, canal cleaning, under-tree irrigation, and desuckering), insect control (weed control, weevil sampling), for nematode management (Nemobiol). The irrigation system used on the farm is under-tree sprinkler irrigation, fed by a deep well (brackish water EC: 1.6 dS/m) and by the Taiguaguai irrigation system in the Aragua state.

The harvest is estimated using the bagging inventory; the bunch pruning is carried out 2 weeks after flowering; the type of bunch pruning used is false +2; the bunch is deflowered; the form is in three banana hands; the acorn removal is carried out when the bunch is 2 weeks old; sponges are used between their hands to avoid damage from rubbing; the bunches are harvested when they are between 13 and 14 weeks old. The bunches are transported to the packing house using a cable track system.

2.4.2.2 Field A2

Field A2 has a cultivated area of 9 ha, and 100% of the production is destined for the local market. Fertilization is carried out by applying complete fertilizers (12 cycles/year) and simple fertilizers (2 cycles/year). Phytosanitary management is carried out by applying systemic fungicides (6 cycles/year); weed control is carried out by applying post-emergent herbicides: paraquat and glyphosate (12 cycles/year); nematode control is carried out using carbamate nematicides (6 cycles/year); to control insects, Chlorpyrifos is applied in covers (12 cycles/year).

The cultural practices used are clearing, weed control for phytosanitary management, pocketing for insect management, and placing leaves in the streets for weed control. The irrigation system applied is by short furrows, fed by deep wells; it does not have a drainage system.

The harvest is estimated through the size of the fruit and the age of the bunch; bunch pruning and acorn removal are carried out when the fruit is 1 week old; the type of bunch pruning applied is false +2; acorn removal is carried out when the bunch is 2 weeks old; the fruit is carried manually using plastic baskets. The harvest is carried out when the fruit is 15 weeks old; they do not determine the length and thickness of the banana using parameters before harvesting.

2.4.2.3 Field A3

It has a cultivated area of 20 ha, and 100% of the production is directed to the local market. Fertilization is carried out by applying complete formula fertilizers (6 cycles/year). Phytosanitary management is done by applying four cycles of systemic fungicides. Weed control is exercised by applying post-emergent herbicides (paraquat and glyphosate) 16 cycles/year; for insect control, Chlorpyrifos is applied in covers (12 cycles/year). Cultural practices are also carried out for phytosanitary management (clearing, topping, weed control), insect management (pocketing), and weed management (leaves are placed in the street).

Irrigation is carried out by a system of short furrows fed by a deep well; there is no network or drainage system. The method used to estimate the harvest is the age of the bunch; leaf removal and weeding are done when the fruit is 1 week old, and the type of bunch pruning applied is false +2. Acorn removal is done when the bunch is 2 weeks old; the fruit is carried manually in plastic baskets, and it is cut when it is 15 weeks old; it is important to point out that they do not determine the length and thickness of the banana before harvesting.

2.4.2.4 Field A4

The farm has a cultivated area of 9 ha, and 100% of the production is directed to the local market. Fertilization is carried out by applying complete formula chemical fertilizers (3 cycles/year). Phytosanitary management is carried out by applying systemic and protectant fungicides (4 cycles/year of systemic fungicides and 2 cycles/year of protectant fungicides); that is, six cycles of fungicides are applied per year. For weed control, post-emergent herbicides such as glyphosate are applied (12 cycles/year). Insect control is carried out by applying Chlorpyrifos (4 cycles/ year); cultural practices are also carried out for disease management (clearing (1 time/month), pruning, weed control (12 times/year), furrow irrigation), insect management (pocketing), and weed management (leaves are placed in the street).

The irrigation system used is by short furrows fed by deep wells; there is no drainage system or network. The harvest is estimated through the national market when the bunch is 15 weeks old; bunch pruning is done when the fruit is 1 week old; the type of bunch pruning used is false +2, and acorn removal is done when the fruit is 1 week old. Acorn removal is done when the fruit is 2 weeks old. The bunch is transported directly to the truck, and there is no packaging. The bunches are harvested when they are 15 weeks old, and the thickness and length of the bananas are not determined before harvesting.

2.4.2.5 Field T1

It has 300 ha of cultivated area, where 92% of the production is directed to the international market and only 8% is destined for the national market. Chemical and organic fertilization is carried out as follows: 6 cycles/year of complete formula chemical fertilizers, 4 cycles/year of foliar fertilizers, 1 cycle/year of organic fertilization through the application of sugar cane compost, and 4 cycles/year of potassium nitrate; that is, 15 cycles/year of mixed fertilization are carried out.

Regarding phytosanitary management for the control of Sigatoka, systemic fungicides and protectants are applied, and weed control is carried out using post-emergent herbicides (5 cycles/year with paraquat, diquat, and glyphosate). For the control of insects, they carry out 52 cycles/year of Chlorpyrifos applied to the sheaths of the bunches.

In addition, on the field, cultural practices are carried out for the management of Sigatoka (weed control, drainage, under-tree irrigation). In the management of insects, traps with soap and water and the bag are used, and for the management of weeds, leaves are placed in the streets. Irrigation is carried out by a sub-tree sprinkler system, which is fed by deep wells; the drainage system is made up of a network of open trapezoidal drainages (primary, secondary, and tertiary channels).

The method to estimate the harvest is the population count and historical harvest data. The type of bunch pruning is done when the fruit is 1 week old through the false +3 technique. Acorn removal is done in the field when the fruit is 1 week old; fruit removal is done on the entire bunch, and acorn removal is done when the bunch is 2 weeks old. For the harvest, sponges are used between the hands to avoid damage due to rubbing; the harvested bunches are transported and carried to the packing plant using a cableway. The bunches are harvested when they are 12 weeks old, and indicators are also used for the thickness and length of the bananas before harvesting.

2.4.2.6 Field T2

It has 270 ha of cultivated area; 20% of the production is directed to the international market, while 80% is destined for the local market. Chemical fertilization (4 cycles/year) is carried out on the farm through the application of ammonium sulfate. It is a farm in the process of recovery.

Phytosanitary management is carried out by applying systemic fungicides (8 cycles/year) and protectant fungicides (1 cycle/year); that is, nine cycles of fungicides are applied per year. For weed control, post-emergent herbicides (paraquat, diquat, and glyphosate) are used (5 cycles/year). For the control of insects, insecticides are applied (52 cycles/year) in covers (Chlorpyrifos). Cultural practices are employed for disease management (bush management, pruning, weed control, drainage, drip irrigation), insect management (pocketing), and weed management (leaves are placed in the street). The irrigation system used is drip, which is fed by deep wells; the drainage system is characterized by a network of open trapezoidal drains (primary, secondary, and tertiary channels).

Harvest is estimated through population counts and historical harvest data; bunch pruning is done when the fruit is 1 week old; the type of bunch pruning used is false +2; weeding is also done at 1 week old; fruit removal is done on the entire bunch; acorn removal is done when the bunch is 2 weeks old.

For the harvest, sponges are used between their hands to avoid damage by rubbing; the bunches are carried to the packer using a cable track, and the cut or harvest is done when the bunch is 14 weeks old; indicators are used to determine the thickness and length of bananas before harvest.

2.4.3 *Soil Sampling*

In each sampled site, soil pits (60 cm wide, 60 cm long, and 60 cm deep) were made to characterize the soils under study. From them, a total of 1080 disturbed and undisturbed soil samples were taken from the six evaluated sites in the first three soil horizons according to the Soil Quality and Health Diagnostic Guide of Rosales et al. (2008). In each plot of the larger fields (A1, T1, and T2), four soil pits were made for each plot and productivity class, accounting for a total of 48 soil pits (3 fields × 4 plots × 4 replications) (Fig. 2.4). In the case of the smaller fields (A2, A3, and A4), two soil pits were made for each plot and productivity class, accounting for a total of 12 soil pits (3 fields × 2 plots × 2 replications) (Fig. 2.4).

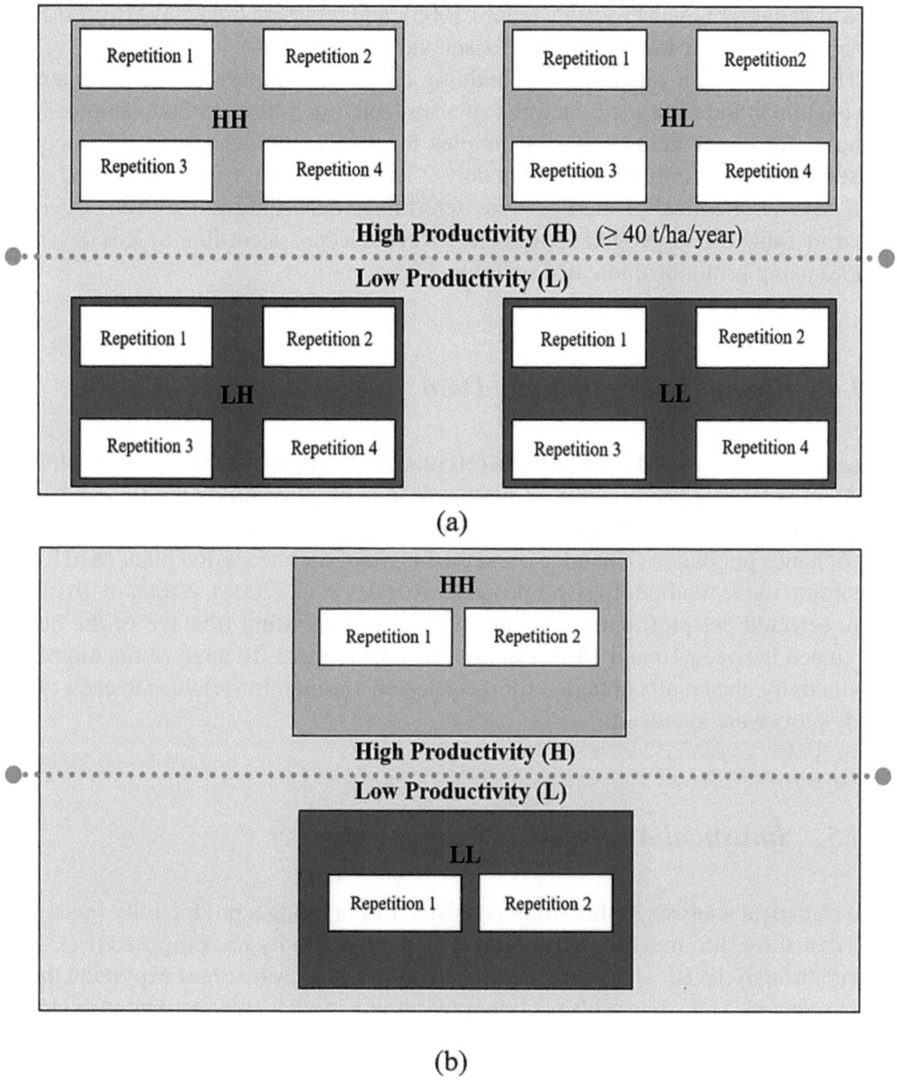

Fig. 2.4 Sampling design for the selection of areas with different banana productivity. (**a**) Farms of size ≥50 ha, (**b**) farms of size ≤25 ha

The undisturbed samples were collected with an Uhland-type sampler, obtaining three replications with an average depth of 29.6 ± 17.6 cm, and there were nine samples for each soil pit. In this way, a total of 540 undisturbed samples (432 samples for the three larger fields and 108 samples for the three smaller fields) were obtained for the determination of each physical property of the soil. These physical properties were bulk density, total porosity, saturated hydraulic conductivity, modulus of rupture according to Pla (1983), resistance to penetration using 150 kPa as a critical value for banana roots (Vaquero 2005), and moisture index (W) expressed as the gravimetric soil water content (Cassel and Nielsen 1986).

The same design was used for obtaining disturbed samples for measurement of the chemical and biological analysis of soils, totaling 540 disturbed samples (432 samples for larger fields and 108 samples for smaller fields) taken and placed in plastic bags and sampling approximately 2 kg of soil. From these disturbed samples, several chemical properties were determined according to the protocols indicated in Table 2.2. Biological parameters were selected according to Rosales et al. (2008) using protocols indicated in Table 2.2.

2.4.4 Biometric Productivity Data

In each of the evaluated banana plots (60 plots in total), four representative plots of 1000 m^2 were selected to estimate biometric data on plant productivity: circumference of the pseudostem of the mother plant at 1 meter in height (CircM, cm), number of hands per bunch (Nhand, n), and the height of the succession plant (AltH, cm) according to the methodology proposed by Rosales et al. (2008). A total of 20 plants were selected per plot, with their bunch close to harvesting (the age of the bunch fluctuated between 10 and 13 weeks after its appearance). To analyze the biometric productivity, the results obtained for the selected agronomic variables in each of the study plots were averaged.

2.4.5 Statistical Analyses

The statistical analysis in this study consisted of obtaining a productivity index (PI) based on three different biometric data on plant productivity was proposed (Fig. 2.5). Then, through the RF algorithm, the first 16 soil variables that best explained the PI were selected. Thus, five multiple linear regression models were estimated using the stepwise regression method. Subsequently, bootstrapping was used for the regression models to check the stability of the method, calculating the confidence intervals of the regression coefficient, and finally, a performance analysis was used to compare the prediction quality range and the error associated with the number of soil variables selected for the proposed models.

2.4.5.1 Productivity Index

A productivity index (PI) was developed based on the biometric production data evaluated in the plots (i.e., CircM, Nhand, and AltH). The developed PI was generated from a principal component analysis (PCA) in which the variables best represented in the first axis (Principal Component 1) were retained. PCA is a technique

Table 2.2 Methods for the determination of the chemical, biological, and root properties of the sampled banana plots (Olivares et al. 2020a, b)

Indicator	Method	Reference
Chemical properties		
pH of soil	McLean method; soil-water ratio, 1:2.5	USDA (1995)
Organic matter (%)	From the oxidizable organic carbon by the Walkley and Black method	USDA (1995)
Electrical conductivity of the soil ($dS.m^{-1}$)	Conductivity meter	USDA (1995)
Nitrate ($cmol.kg^{-1}$)	Extractor solution KCl, K_2SO_4	Brinford et al. (1992)
Calcium carbonate ($cmol.kg^{-1}$)	Horton and Newsom's method	Horton and Newsom (1953)
Aluminum, exchangeable acidity, calcium, magnesium, potassium, sodium, phosphorus, iron, copper, zinc, manganese ($cmol.kg^{-1}$)	Extracted with Mehlich 3	Mehlich (1984)
Biological properties		
Total population count of bacteria, fungi, and actinomycetes (colony-forming units. g^{-1} soil)	Direct counting techniques	Weaver et al. (1994)
Microbial respiration (mg 100. g^{-1}.10 $days^{-1}$)	CO_2 release	Alef (1995)
Microbial biomass (mg)	The fumigation-extraction method with chloroform	Joergensen (1995)
Carbon associated with microbial biomass (mg)	(mg C in 100 g dry soil at 105 °C; the flow of C in flow-C / factor 0.45; fumigated and not fumigated)	Joergensen (1995)
Radical weight (functional, nonfunctional, and total roots) (g)	Total extraction of plant roots	Bongers (1990)
Plant-parasitic nematodes: *Radopholus similis* and *Helicotylenchus multicinctus* (logarithm (N + 1))	Nematode extraction from plant roots	Bongers (1990)
Total free-living nematodes (number of nematodes in 100 g root)	Modified Baermann's method	Bongers (1990)
Number of genera of free-living nematodes		

for the construction of multivariate indices used in different areas of knowledge, e.g., Primpas et al. (2010) and Araya-Alman et al. (2019). The objective of PCA is to reduce the dimensionality of the dataset. In addition, the linear combination of the three biometric variables evaluated was used as a synthesis variable for the analysis using the XLSTAT software (Addinsoft 2010) with the coefficients of this linear combination being the loadings of the variables in PC1.

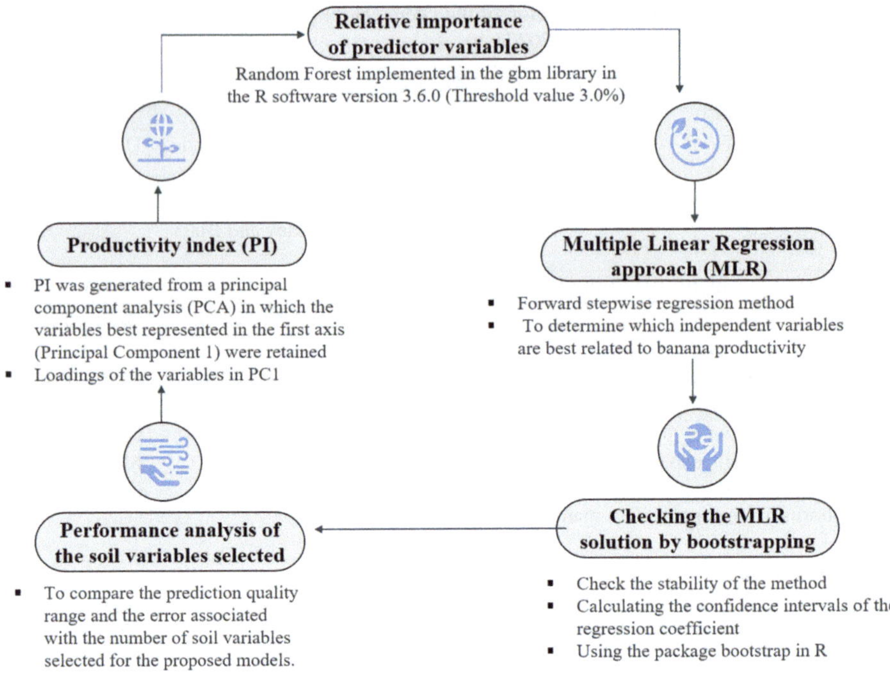

Fig. 2.5 General scheme of the data analysis procedures

2.4.5.2 Relative Importance of Predictor Variables

Random forest (RF) is a technique that allows selecting the most important predictors; however, the most usual procedure in forward stepwise estimation is the analysis of the correlation matrix and multicollinearity testing (e.g., analysis of tolerance or variance inflation factor, VIF). This random forest technique allows for reducing multicollinearity problems between variables. A high degree of multicollinearity can have substantive effects on the regression coefficients estimation and their statistical significance.

An RF is a collection of predictor trees $h(\mathbf{x}; \theta_k)$, $k = 1 \ldots K$ where $\mathbf{x}$ represents the observed input vector (covariate) of length p with associated random vector $\mathbf{x}$ and θ_k are independent and identically distributed (*iid*) random vectors. This type of model optimizes the variables selected in the analysis according to the percentage of contribution of the variable response (Breiman et al. 2003).

The relative importance of the soil variables used in the study through the RF was obtained using the equations developed by Friedman (2001) and implemented in the *gbm* library in the R software version 4.0.2 (R Core Team 2020). The measures are based on the number of times a variable is selected for splitting, weighted by the squared improvement to the model because of each split, and averaged over all trees (Friedman and Meulman 2003). This relative importance (or contribution) of each variable is scaled so that the sum adds up to 100, with higher numbers indicating a greater influence on the response; in this case, a threshold value of 3.0% was used (Elith et al. 2008).

2.4.5.3 Multiple Linear Regression (MLR) Approach

The soil variables selected through random forest were subjected to an automatic selection of predictive variables to determine which independent variables are best related to banana productivity. For this reason, a multiple linear regression model was performed using the forward stepwise regression method, as proposed by Miller (2002). The model used for multiple linear regression is presented in Eq. 2.2 using the *lm* function in Rpackagelme4.

$$Y_i = \beta_0 + \beta_1 X_1 + \beta_2 X_2 + \beta_3 X_3 + \cdots + \beta_k X_i + \varepsilon \quad (2.2)$$

where Y_i is the banana productivity index (PI), β_0 is the intercept, β_1 to β_k are the model regression parameters, X_1 to X_i are the soil variables, and ε represents the random error.

2.4.5.4 Checking the MLR Solution by Bootstrapping

Bootstrapping was applied to the regression models to check the stability of the method, calculating the confidence intervals of the regression coefficient using the package bootstrap version 2019.6 in R. The model was considered stable when 95% of the values of the sampled coefficients are within the range (lower-upper) of the estimated confidence intervals. This would demonstrate that the method is stable and generalizable, saving the tedious verification of all the basic assumptions of the regression (normality, homogeneity of variances, etc.) and having the certainty that overfitting has been avoided without having to resort to the complex cross validation (Efron and Tibshirani 1993).

2.4.5.5 Performance Analysis of the Soil Variables Selected

A performance analysis was used to compare the quality range of prediction and the associated error in the number of soil variables selected. In the first case, a specific number of variables were used to estimate the PI that was related to different multiple linear regression models. In the second case, different error rates were calculated to obtain the optimal PI model. Then, the performance analysis considered three statistical measurements to assess the accuracy of the models finally proposed: (i) the root mean squared error (RMSE), (ii) the mean absolute error (MAE), and (iii) the coefficient of determination (R^2) (Acevedo-Opazo et al. 2013).

2.5 Study of Main Categorical Soil Properties with the Banana Productivity

Soil morphological indicators are dynamic indicators, capable of changing under different management regimes and land use pressures, being sensitive to change; they quickly warn of changes in soil conditions and constitute an effective monitoring tool. Field soil morphological properties are among the most important soil properties for genesis and classification. They can be found easily and cheaply, are included in virtually all soil databases, and can be easily applied by farmers.

However, as they are not numerical, their quantitative use in soil quality creates significant difficulties (MacEwan and Fitzpatrick 1996; Calero et al. 2018). Thus, there are few antecedents for the calculation of soil quality indices based on pedomorphological indicators, which were established solely through knowledge-based decision rules (Onweremadu et al. 2008; Pulido-Moncada et al. 2014).

Soil morphological variables, expressed in qualitative scales (ordinal or nominal), can be an interesting option when it comes to grouping and/or predicting soil fertility for crops, given the simplicity with which they are obtained. For a correct and rigorous treatment of the morphological variables using models, especially of a multivariate type such as factorial analysis or multiple regression, these must be transformed to a numerical scale. The optimal scaling procedures allow this transformation to be carried out using functions that assign numerical values to the categories of the morphological variables increasing the explanatory capacity of the models (% variance explained in principal component analysis or R^2 in regression analysis) (Olivares 2016; Olivares et al. 2017a, b).

In what follows, this study will proceed to apply two types of multivariate models with optimal scaling—first, an exploratory categorical principal component analysis (CATPCA) (Calero et al. 2005, 2008) and second, a categorical regression (CATREG) (Yang et al. 2020)—to the productivity data of banana plantations in Venezuela. Using CATPCA, the dimensionality reduction of the initial set of morphological variables and its transformation are achieved, so numerical statistical tests can be applied to verify the ability to discriminate between production levels (high and low) of the soil morphological variables and the components. Using CATREG, it is intended to develop a quantitative regression model between a quantitative index, the banana production index (PI) previously developed based on the morphometric characteristics of the plants (for each plot) (see Sect. 2.4.4), and the morphological variables of the soil. This study is based on the correlation of banana productivity levels and soil morphological properties using regularized optimal scaling regression.

2.5.1 Study Areas

The same six banana plantations in Chap. 5, in the states of Aragua and Trujillo in Venezuela, were selected. The plantations in the state of Aragua (PL, SM, PZ, and CH) are in the Lake Valencia Basin. Plantation PL is located on the fourth level of the lacustrine terrace produced by drying of Lake Valencia, while plantations SM, PZ, and CH are on alluvial soils, all with medium to silty textures. Soils in these farms are Mollisols and Inceptisols and are fertile, generally with moderate to good drainage, soil pH varying from neutral to alkaline, and medium to high organic matter content (Delgado et al. 2010).

The second study area is in the state of Trujillo (plantations BA and KA), in the region southeast of Lake Maracaibo, also characterized by a Tropical Savanna (Aw) climate. This area of the state of Trujillo is an alluvial plain with slopes of less than 1.0% with mainly Entisol soils (Rodríguez et al. 2006). These soils have moderate to poor drainage with neutral to alkaline pH. They are moderately fertile, and the average organic matter content is around 2.75% (Rey et al. 2009). In both areas, soil management was concentrated on fertilization, and no reclamation action was taken to improve drainage or increase organic matter.

2.5.2 Soil Sampling

Banana productivity was estimated in sampling areas delimited by productivity level following the guidelines proposed by Rosales et al. (2008). On all the plantations, two plots or productivity levels were identified a priori as High (H) and Low (L) for estimating the productivity index (PI) (Fig. 2.6). The "Gran Nain" variety was the only variety grown, and each productivity plot had an area of 4 ha (on the large >50 ha plantations, PL, BA, and KA) with four replicated plots in each field. On large plantations (≥50 ha, PL, BA, and KA), the average yield of high-productivity plots was 69.8 ± 5.0 t ha^{-1} $year^{-1}$, and in low-productivity plots, it was 59.7 ± 5.3 t ha^{-1} $year^{-1}$. On the remaining small plantations (<25 ha, SM, PZ, and CH), these two levels of productivity were identified in 1-ha fields, with two replicated plots in each field, for a total of 36 plots. The average yield on small plantations was 11.5 ± 0.7 t ha^{-1} $year^{-1}$ for high-productivity plots and 1.6 t ha^{-1} $year^{-1}$ for those with low productivity.

2.5.3 Productivity Index (PI)

The study used the banana productivity index (PI) previously developed by Olivares et al. (2020a, b) to estimate productivity in each of the evaluation plots. The PI is based on the morphometric characteristics of banana plants, such as the circumference of the mother plant pseudostem at 1 m height (cm), the number of hands per bunch (n), and height of the succession plant (cm), following the methodology

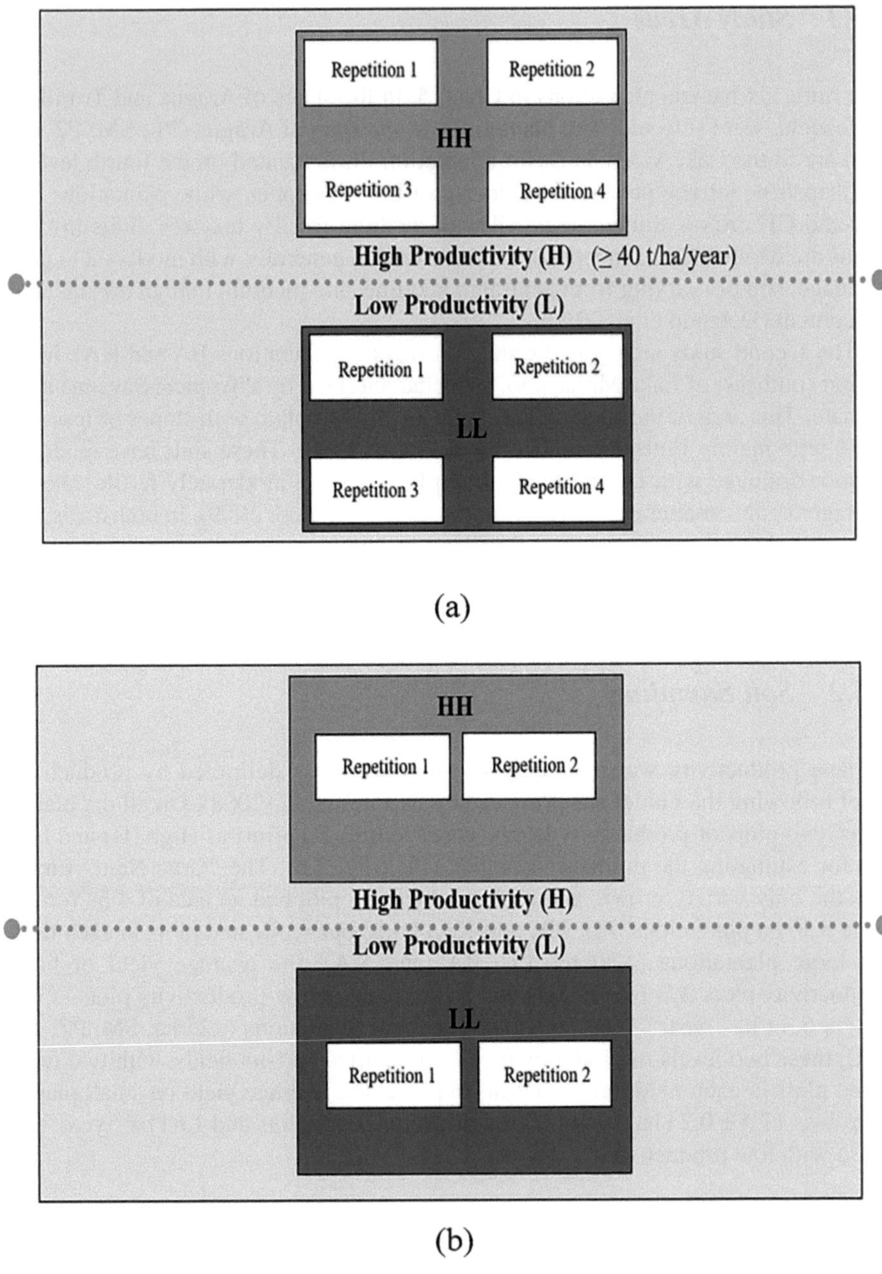

Fig. 2.6 Sampling design for the selection of areas with different banana productivity. (**a**) Farms of size ≥50 ha, (**b**) farms of size ≤25 ha

proposed by Rosales et al. (2008). The PI was generated from a principal component analysis (PCA) in which the variables best represented on the first axis (Principal Component 1) were retained. In addition, the linear combination of the three biometric variables evaluated was used as a synthesis variable, where the coefficients of this linear combination were the variable loadings in PC1. Then, in the categorical regression, the ordinal scaling level was considered the response variable.

2.5.4 *Morphological Characterization of Soil Properties*

A soil profile was evaluated in each of the 36 replicated plots indicated in Sect. 2.2. The A horizon morphological properties were described and characterized in the field following the FAO (2006) and the Soil Survey Staff (2017) methodologies. The color was determined according to the Munsell soil color charts (Munsell Color Company 1999). The texture was determined in the laboratory by sieving and sedimentation using a Robinson pipette (Soil Survey Staff 2017). Seventeen morphological variables and the horizon thickness in these soil profile descriptions were studied: texture class, structure size, structure grade, structure type, moist hue, moist value, moist chroma, moist consistence, dry consistence, stickiness, plasticity, biological activity, root abundance, root size, cutans thick, cutans frequency, and reaction to HCl.

2.5.5 *Statistical Modelling of Field Data*

2.5.5.1 Categorical Principal Component Analysis (CATPCA)

First, the CATPCA was applied as an exploratory analysis to transform by optimal scaling the categorical variables (ordinal, nominal) to numerical ones and compare the means of each production level (H, L) using Student's t test; in this way, it will be possible to establish which variables are associated with higher productivity in bananas. This exploratory analysis was first performed at the horizon level ($N = 208$). Since, contrary to the standard PCA, a specific number of components must be indicated to the model, the variance explained by each component (eigenvalue) must be explored. The selection of the number of components is done through a scree plot analysis based on the "Kaiser rule" (Linting and van Der Kooij 2012).

The 17 morphological variables have been introduced into the model together with the thickness of each horizon (total of 18 variables). The measurement scale defined for each variable is described in Table 2.3. The variables texture, hue, and type of structure were assigned a nominal scale since their categories do not present intrinsic ordering (it is true that hue presents an intrinsic ordering, but its scale is "circular," so it is preferred not to force its scaling).

Table 2.3 Type of morphological variable and measurement scale (numerical, nominal, and ordinal)

n	Variable	Measurement scale
1	Thickness	Numerical (discretized)
2	Texture	Nominal
3	Hue	Nominal
4	Value	Ordinal
5	Chroma	Ordinal
6	Structure type	Nominal
7	Structure size	Ordinal
8	Structure grade	Ordinal
9	Dry consistence	Ordinal
10	Moist consistence	Ordinal
11	Stickiness	Ordinal
12	Plasticity	Ordinal
13	Biological activity	Ordinal
14	Root abundance	Ordinal
15	Root size	Ordinal
16	Cutans thickness	Ordinal
17	Reaction to HCl	Ordinal
18	Cutans frequency	Ordinal

Once the morphological variables have been transformed into numerical values, the statistical correlation with some more precise production indicators can be studied in more detail than the simple classification of banana soils into groups of high or low productivity. In this case, Olivares et al. (2020a, b) developed a quantitative index (PI) for each plot that they use as a response variable in a regression analysis based on soil analytical variables (physical and chemical). The index is developed for the same soils from which the morphological variables have been obtained (discarding two intermediate productivity levels, HL and LH), although from a systematic sampling at a depth of 60 cm in which no took into account the distribution of profile horizons (N = 208). The correlation with the transformed morphological properties, therefore, will also be checked against the average weighted by profile up to its total depth and up to a depth of 60 cm (N = 36); the latter would allow for establishing relationships with analytical properties.

2.5.5.2 Regression with Optimal Scaling

First, a regression model was fitted to be able to predict the banana productivity index (PI) (*outcome*) based on the soil morphological properties (*predictors*) on the plantations sampled. However, the categorical or nonmetric nature of the predictors prevents the proper application of classical regression analysis. Various techniques have been applied to include categorical variables in multiple regression, such as the creation of dummy variables. These variables, however, introduce high multicollinearity and hinder the interpretation of the model, reducing the significance of the regression coefficients (Wissmann et al. 2011) and the predictive power of the model (Hair et al. 1999; Xu et al. 2010).

The problem of multicollinearity between predictors can be dealt with by selecting a good theoretical model, which enters only those predictors with a high degree of unique variance with the dependent variable (Hair et al. 1999). However, it is hard to select categorical predictors using the usual stepwise procedure, where each variable is selected or discarded before its transformation to a numeric scale. For rigorous treatment of categorical variables, such as field morphological variables, in multivariate methods, optimal scaling methods need to be applied (Calero et al. 2018).

Our analysis was applied to the A horizons ($N = 63$) in each soil. The average thickness of these horizons is 31 ± 12 cm, which roughly coincides with the area of maximum concentration of the banana tree (30 cm). We hypothesized that focusing our analysis on the "A" horizon would make more agronomic sense since this is the area that most influences plant development and yield. The functional roots of the banana in our profile descriptions were concentrated in the A horizons, which ranged from about the top 30 to 40 cm deep, a root concentration range like the description of banana in the tropics by Gauggel et al. (2003).

The formal statistical analysis in this study consisted of the implementation of the categorical regression (CATREG) (Fig. 2.7), which is an optimal scaling procedure in which the morphological variables are transformed into a numerical scale and can thus be entered in a multiple regression analysis. In CATREG, the goal is to minimize the dataset prediction error. However, in this case, there is no test dataset, and therefore, two resampling methods were used: k-fold cross validation (CV) and the 0.632 Bootstrap, using the mean square error (MSE) for assessing the expected prediction error (EPE). Next, regularization is used for selecting the model and avoiding overfitting in predictive techniques. The regularization technique Lasso is one of the most widely employed for optimal scaling regression.

Categorical regression (CATREG) (van der Kooij et al. 2006; Meulman et al. 2019) is an optimal scaling technique that can transform the k_j categories ($s = 1, \ldots,$

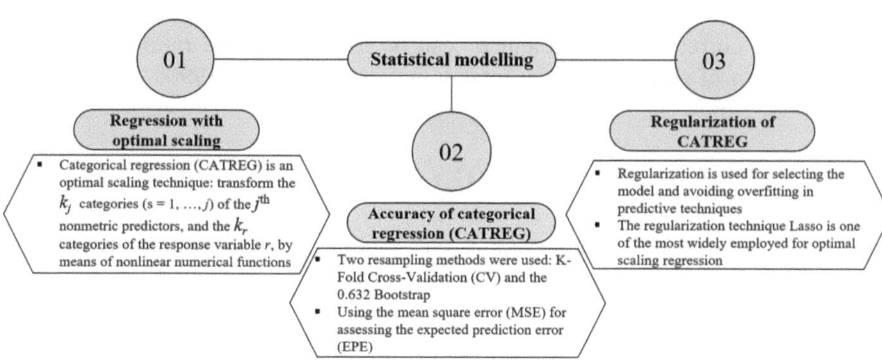

Fig. 2.7 General scheme of the data analysis procedures

j) of the *j*th nonmetric predictors and the k_r categories of the response variable *r* using nonlinear numerical functions while minimizing the error (i.e., increasing the coefficient of determination R^2 of the model). CATREG takes the classical linear regression model with nonlinear optimal scaling (Eq. 2.3):

$$\varphi(y) = \sum_{j=1}^{p} \beta_j \varphi(x) + e \tag{2.3}$$

where β_j are the regression coefficients of the *j*th predictor, $\varphi(x)$ and $\varphi(y)$ are the transformation functions of predictors (x) and response variable (y), respectively, and e is the model error. In this study, the only restriction applied to $\varphi(x)$ and $\varphi(y)$ was monotonicity, which makes it possible to distinguish between field soil morphological properties measured on an ordinal scale (having an intrinsic categorical order, i.e., stickiness: from not sticky to very sticky) from nominal variables not having this restriction. Hue, structure type, and texture class were the second types. Rewriting Eq. 2.3 in terms of indicator matrices and category quantifications yields Eq. 2.4:

$$\boldsymbol{G}_r \boldsymbol{y}_r = \sum_{j=1}^{p} \beta_j \boldsymbol{G}_j \boldsymbol{y}_j + e \tag{2.4}$$

where $\boldsymbol{\beta}$ ($\beta_1, \ldots, \beta_p$), in order p (the number of predictor variables), is the vector containing regression coefficients, $\boldsymbol{y}_j$ and $\boldsymbol{y}_r$, in order k_j and k_r, the optimal scaling's or numerical transformations of the categories for the predictors and response variable, respectively, and $\boldsymbol{G}_j$ and $\boldsymbol{G}_r$, in order $n \times k_j$ and $n \times k_r$ (where n is the number of cases), indicator matrices, such that 1 is when the *i*th object is in the k_j category of variable *j* and 0 otherwise. CATREG estimates the regression coefficients by minimizing the least squares loss function (van der Kooij et al. 2006) in Eq. 2.5.

$$\sigma\left(\boldsymbol{y}_r; \beta; \boldsymbol{y}_j\right) = \boldsymbol{G}_r \boldsymbol{y}_r - \sum_{j=1}^{p} \beta_j \boldsymbol{G}_j \boldsymbol{y}_j^{\,2} \tag{2.5}$$

The multiple correlation coefficient R^2 can be found from the ratio between the regression sum of squares and the total sum of squares (Gifi 1990; van der Kooij 2007) in Eq. 2.6:

$$R^2 = N^{-1/2} \left(\boldsymbol{G}_r \boldsymbol{y}_r\right)' \mathbf{v} \left(\mathbf{v}\mathbf{v}'\right)^{-1/2} \tag{2.6}$$

where N is the number of observations and $\mathbf{v}$ is the accumulated contribution of predictor variables so that

$$\mathbf{v} = \sum_{j=1}^{p} \beta_j \boldsymbol{G}_j \boldsymbol{y}_j \tag{2.7}$$

The statistics (t, F values) and fit and error measures, as well as the correlation matrices R, partial correlation, and predictor tolerance, which will be used to assess the goodness of the model, are found from β_j and R^2. Tolerance is defined as one less the determination coefficient R^2 of the prediction of any predictor by the others, considering their independent variables, so it should be high to avoid multicollinearity.

2.5.5.3 Accuracy of Categorical Regression (CATREG)

The error term in the regression model $(1 - R^2)$ is usually not a good estimator of the prediction error, since it is found from the same data used to fit the model. To obtain a better estimate, resampling methods, such as cross validation or bootstrapping, may be used.

In CATREG, the goal is to minimize the dataset prediction error. The expected prediction error (EPE) must be known to be able to assess the reliability of predictions of future observations (van der Kooij 2007). Therefore, ideally, a training dataset would be used to estimate a model and a test dataset. However, in this case, there is no test dataset, and therefore, two resampling methods were used: k-fold cross validation (CV) and the 0.632 Bootstrap, using the mean square error (MSE) for assessing the EPE.

In k-fold cross validation, data are divided into randomized k groups of approximately the same size. $K - 1$ groups are used to train the model, and another is used for validation. This process is repeated k times using a different group as validation in each iteration. The process generates k estimates of the error that are averaged as the final estimate (Meulman et al. 2019). In bootstrapping, N cases of the same size as the sample are taken at random from the full dataset in each resampling to obtain B bootstrapped samples. Then, a model is fitted for each bootstrapped sample, estimating its prediction error from the original (not resampled) dataset. Contrary to CV, random sampling is performed with replacement; that is, an observation can be repeated in the bootstrapped sample, while another is excluded (*out-of-bag* set). The simple bootstrap estimate of the expected prediction error is found by averaging the error estimates of the B bootstrapped samples. Since the bootstrapped (training set) and the original sample (validation set) have many observations in common, the bias seems to be greater than with CV, and in general, it performs worse. Efron (1983) proposed a modification of the bootstrap prediction error called the 0.632 method, which includes the bootstrapped model error estimates in the simple bootstrap in their *out-of-bag* observations (not the original dataset). This corrects the bias and improves model performance (van der Kooij 2007). Here, we used ten folds for CV and 50 bootstrap samples in 0.632 bootstrapping. In addition to the reliability of the prediction estimated by the MSE, a more flexible approach can be used to evaluate the CATREG model accuracy (Hartmann et al. 2009). For this, the dependent variable (PI, which usually ranged from −4 to +4) was distributed into two classes or groups: N, negative with PI <0, and P, positive with PI >0, using zero as the cut-off point. Thus, model accuracy could be addressed by such classification accuracy measures as the efficiency and the receiver operating characteristic (ROC) curve (Garosi et al. 2019). Efficiency can therefore be defined in Eq. 2.8.

$$\text{Efficiency} = \frac{TP + TN}{TP + TN + FP + FN} \tag{2.8}$$

where TP (true positive) and TN (true negative) represent the number of correctly classified P and N plots, respectively, FP (false positive) is the number of N plots that have been classified as P, and FN (false negative) those P considered N.

The ROC curve plots the sensitivity (the ratio of the number of correctly classified P to total observed P) versus "1 – specificity" (specificity is the ratio between the number of correctly classified N and the total observed N). A model's predictive performance is high if high sensitivity is obtained at low values of "1 – specificity," that is, good capacity for correctly classifying P with a low number of false positives. This yields a curve closer to the upper left-hand corner (Garosi et al. 2019). The area under the ROC curve (AUC) quantifies this relationship, so a model is considered acceptable if AUC ≥0.7, excellent if AUC ≥0.8, and outstanding if AUC ≥0.9.

2.5.5.4 Regularization of CATREG

Regularization is used for selecting the model and avoiding overfitting in predictive techniques, since the estimation of the regression coefficients by least squares may present collinearity (James et al. 2013). It is especially useful in categorical regression (Meulman et al. 2019). Of the three main regularization techniques, Ridge, Lasso, and elastic net, Lasso is one of the most widely employed for optimal scaling regression. As first developed by Tibshirani (1996), Lasso regularization can deal with complex models with many predictors and high multicollinearity, when ordinary least squares show instability and overfitting. Both effects (instability and overfitting) are common in optimal scaling regression and can make it difficult to select a satisfactory theoretical model (Hartmann et al. 2009).

Lasso applies a λ penalty to the CATREG loss function that reduces the estimated regression coefficients, shrinking them to zero as the penalty increases (van der Kooij 2007) in Eq. 2.9.

$$\sigma^{\text{lasso}}(\beta) = \boldsymbol{G}_y \boldsymbol{y}_r - \sum_{j=1}^{p} \beta_j \boldsymbol{G}_j \boldsymbol{y}_j{}^2 + \lambda \sum_{j=1}^{p} \left|\beta_j\right| \tag{2.9}$$

Predictors with the most stable estimate of the coefficient shrink to zero more slowly, so Lasso regularization can be used to advantage of exploratory analysis instead of stepwise procedures to obtain a set of predictors with low multicollinearity. The advantage of Lasso over alternatives like Ridge regularization is that it shrinks the coefficients to zero, which the others do not. By canceling coefficients, model interpretation (coefficient selection) becomes straightforward. Lastly, the elastic net regularization combines the Ridge penalty rule and Lasso but involves a complicated calculation, and its interpretation is not as immediate as Lasso's. For details on the mathematics of these three regularization techniques in optimal scaling regression, see Meulman et al. (2019).

Since each λ penalty involves a regression model, the model that the regularized regression coefficients will be based on must be selected. We followed the procedure specified by Meulman et al. (2019) for this, selecting the most parsimonious model with the lowest prediction error. As an increase in the penalty usually yields lower prediction errors, the best model would be the one in which most of the coefficients are canceled out, that is, when the sum of the standardized regression coefficients is very close to zero. However, a local minimum might be selected, such that it keeps a suitable number of coefficients, enabling an adequate theoretical model to be established. The one standard error (*1-SE*) rule can be applied for this (James et al. 2013). We chose the regularized model with the lowest number of predictors (the more parsimonious model) within one standard error of the model with the lowest EPE (the optimal model). All statistical analyses, including the regularized regression, were performed with IBM SPSS 24 (IBM Corp 2016).

References

Acevedo-Opazo C, Valdés-Gómez H, Taylor JA, Avalo A, Verdugo-Vásquez N, Araya M, Jara-Rojas F, Tisseyre B (2013) Assessment of an empirical spatial prediction model of vine water status for irrigation management in a grapevine field. Agric Water Manag 124:58–68. https://doi.org/10.1016/j.agwat.2013.03.018

Addinsoft TM (2010) XLSTAT software version 5.03. Addinsoft. Barcelona. Spain. http://www.xlstat.es

Alef K (1995) Field methods. In: Alef K, Nannipieri P (eds) Methods in applied soil microbiology and biochemistry. Academic Press, London, pp 464–470

Araya-Alman M, Leroux C, Acevedo-Opazo C, Guillaume S, Valdés-Gómez H, Verdugo-Vásquez N, Pañitrur-De la Fuente C, Tisseyre B (2019) A new localized sampling method to improve grape yield estimation of the current season using yield historical data. Precis Agric 20:445–459. https://doi.org/10.1007/s11119-019-09644-y

Bentley S, Pegg KG, Moore NY, Davis RD, Buddenhagen IW (1998) Genetic variation among vegetative compatibility groups of *Fusarium oxysporum* f. sp. *cubense* analyzed by DNA fingerprinting. Phytopathology 88:1283–1293. https://doi.org/10.1094/phyto.1998.88.12.1283

Bongers T (1990) The maturity index: an ecological measure of environmental disturbance based on nematode species composition. Oecologia 83:14–19. https://doi.org/10.1007/BF00324627

Bosman M (2016) Role of the environment on the incidence of Panama disease in bananas. Dissertation, Soil Geography, and Landscape, Wageningen University

Breiman L, Last M, Rice J (2003) Random forests: finding quasars. In: Statistical challenges in astronomy. Springer, New York. https://doi.org/10.1007/0-387-21529-8_16

Brinford G, Blackmer A, Cerrato M (1992) Relationship between corn yields and soil nitrate late spring. Agron J 84:59–63. https://doi.org/10.2134/agronj1992.00021962008400010012x

Bubici G, Kaushal M, Prigigallo MI, Cabanas CGL, Mercado-Blanco J (2019) Biological control agents against fusarium wilt of banana. Front Microbiol 10(1). https://doi.org/10.3389/fmicb.2019.01290

Bylesjö M, Rantalainen M, Cloarec O, Nicholson JK, Holmes E, Trygg J (2006) OPLS discriminant analysis: combining the strengths of PLS-DA and SIMCA classification. J Chemom 20(8–10):341–351. https://doi.org/10.1002/cem.1006

CABI/EPPO (2015) *Fusarium oxysporum* f. sp. *cubense*. In: Distribution maps of plant diseases, Map 31, 7th edn. CABI, Wallingford

Calero J, Serrano JM, Aranda V, Sánchez D, Vila MA, Delgado G (2005) Analysis and characterization of olive tree cultivation system in Granada Province (South of Spain) with optimal scaling and multivariate techniques. Agrochimica 49(3):118–131

Calero J, Delgado R, Delgado G, Martín-García JM (2008) Transformation of categorical field soil morphological properties into numerical properties for the study of chronosequences. Geoderma 145(3–4):278–287. https://doi.org/10.1016/j.geoderma.2008.03.022

Calero J, Aranda V, Montejo-Raez A, Martin-Garcia JM (2018) A new soil quality index based on morpho-pedological indicators as a site-specific web service applied to olive groves in the Province of Jaen (South Spain). Comput Electron Agric 146:66–76. https://doi.org/10.1016/j.compag.2018.01.016

Cassel DK, Nielsen DR (1986) Field capacity and available water capacity. In: Methods of soil analysis: part 1—physical and mineralogical methods. American Society of Agronomy-Soil Science Society of America, Madison. https://doi.org/10.2136/sssabookser5.1.2ed.c36

Chawla NV, Bowyer KW, Hall LO, Kegelmeyer WP (2002) SMOTE: synthetic minority over-sampling technique. J Artif Intell Res 16(1):321–357. https://doi.org/10.1613/jair.953

Chong J, Wishart DS, Xia J (2019) Using MetaboAnalyst 4.0 for comprehensive and integrative metabolomics data analysis. Curr Protoc Bioinf 68(1):e86. https://doi.org/10.1002/cpbi.86

Davis R (2005) Fusarium wilt (Panama disease) of banana. Pest Adv Leaflet 42:1–8

de Oliveira Aparecido LE, de Souza RG, da Silva Cabral De Moraes J, Costa CT, de Souza PS (2020) Machine learning algorithms for forecasting the incidence of Coffea arabica pests and diseases. Int J Biometeorol 64(4):671–688. https://doi.org/10.1007/s00484-019-01856-1

Deacon JW, Herbert JA, Dames J (1985) False Panama disorder of bananas. In: Citrus and Subtropical Fruit Research Institute Information Bulletin. pp. 15–18

Delgado E, Trejos J, Villalobos M, Martínez G, Lobo D, Rey J, Rodríguez G, Rosales FE, Pocasangre LE (2010) Determination of a soil quality and health index for banana plantations in Venezuela. Interciencia 35(12):927–933

Dita M, Barquero M, Heck D, Mizubuti ESG, Staver CP (2018) Fusarium wilt of banana: current knowledge on epidemiology and research needs toward sustainable disease management. Front Plant Sci 9. https://doi.org/10.3389/fpls.2018.01468

Dumanski J, Gameda S, Pieri C (1998) Indicators of land quality and sustainable land management. The World Bank, Washington, DC

Efron B (1983) Estimating the error rate of a prediction rule: improvement on cross-validation. J Am Stat Assoc 78:316–331. https://doi.org/10.2307/2288636

Efron B, Tibshirani RJ (1993) An Introduction to the Bootstrap. Chapman & Hall, New York

Elith J, Leathwick JR, Hastie T (2008) A working guide to boosted regression trees. J Anim Ecol 77:802–813. https://doi.org/10.1111/j.1365-2656.2008.01390.x

EPPO (2021) PQR database. Paris, France: European and Mediterranean Plant Protection Organization. https://gd.eppo.int/taxon/FUSACB/distribution. Accessed 20 Feb 2020

Espejo B, Martinez-Guanter J, Pérez-Ruiz M, Lopez-Pellicer FJ, Zarazaga-Soria FJ (2018) Machine learning for automatic rule classification of agricultural regulations: a case study in Spain. Comput Electron Agric 150:343–352. https://doi.org/10.1016/j.compag.2018.05.007

Estabrooks A, Jo T, Japkowicz N (2004) A multiple resampling method for learning from imbalanced data sets. Comput Intell 20(1):18–36. https://doi.org/10.1111/j.0824-7935.2004.t01-1-00228.x

FAO (2006) Guidelines for soil profile description, 4th edn. FAO, Roma

FAO (2021) Marco estratégico para 2022–2030. 42 período de sesiones C2021/7 http://www-faoorg/3/ne577es/ne577espdf Accessed 21 Sept 2022

Feuerman M, Miller AR (2008) Relationships between statistical measures of agreement: sensitivity, specificity and kappa. J Eval Clin Pract 14(5):930–933. https://doi.org/10.1111/j.1365-2753.2008.00984.x

Friedman JH (2001) Greedy function approximation: a gradient boosting machine. Ann Stat 29:1189–1232. https://doi.org/10.1214/aos/1013203451

Friedman JH, Meulman JJ (2003) Multiple additive regression trees with application in epidemiology. Stat Med 22:1365–1381. https://doi.org/10.1002/sim.1501

Garosi Y, Sheklabadi M, Conoscenti C, Pourghasemi HR, Van Oost K (2019) Assessing the performance of GIS- based machine learning models with different accuracy measures for determining susceptibility to gully erosion. Sci Total Environ 664:1117–1132. https://doi.org/10.1016/j.scitotenv.2019.02.093

Gauggel CA, Sierra F, Arévalo G (2003) The problems of banana root deterioration and their impact on production: production experience in Latin America. In: Turner DW, Rosales FE (eds). Banana Root System: towards a better understanding for its productive management: proceedings of an International Symposium/ International Network for the Improvement of Banana and Plantain, Montpellier, France. https://n9.cl/b3wi8

Gee GW, Or D (2002) Particle-size analysis. In: Methods of soil analysis: part 4 physical methods. Soil Science Society of America, pp 255–293. https://doi.org/10.2136/sssabookser5.4.c12

Ghag SB, Shekhawat UKS, Ganapathi TR (2015) Fusarium wilt of banana: biology, epidemiology and management. Int J Pest Manag 61:250–263. https://doi.org/10.1080/09670874.2015.1043972

Gifi A (1990) Nonlinear multivariate analysis. Wiley, Chichester

Guo H, Li Y, Shang J, Mingyun G, Yuanyue H, Bing G (2017) Learning from class-imbalanced data: review of methods and applications. Expert Syst Appl 73:220–239. https://doi.org/10.1016/j.eswa.2016.12.035

Hair JF, Anderson RE, Tatham RL, Black WC (1999) Análisis multivariante. 5° ed. Prentice Hall, Madrid

Hartmann A, Zeeck A, van der Kooij AJ (2009) Severity of Bulimia Nervosa. Measurement and classification into health or pathology. Psychopathology 42:22–31. https://doi.org/10.1159/000173700

Heanes DL (1984) Determination of total organic-C in soils by an improved chromic acid digestion and spectrophotometric procedure. Commun Soil Sci Plant Anal 15(10):1191–1213. https://doi.org/10.1080/00103628409367551

Horton JH, Newsom DW (1953) A rapid gas evolution method for calcium carbonate equivalent in liming materials. Soil Sci Soc Am J 17:414–415

IBM Corp (2016) IBM SPSS statistics for Windows, version 24.0. IBM Corp, Armonk

IPPC (2021) Adopted Standards (ISPMs). Adopted Standards (ISPMs) – International Plant Protection Convention (ippc.int)

James G, Witten D, Hastie T, Tibshirani R (2013) An introduction to statistical learning, vol 112. Springer. https://doi.org/10.1007/978-1-4614-7138-7

Joergensen RG (1995) Microbial biomass. In: Nannipieri AK (ed) Methods in applied soil microbiology and biochemistry. Academic Press, London, pp 375–417

Karger DN, Conrad O, Böhner J, Kawohl T, Kreft H, Soria-Auza RW, Zimmermann NE, Linder HP, Kessler M (2017) Climatologies at high resolution for the earth's land surface areas. Sci Data 4:170122. https://doi.org/10.1038/sdata.2017.122

Kema GHJ, Drenth A, Dita M, Jansen K, Vellema S, Stoorvogel JJ (2021) Editorial: Fusarium wilt of banana, a recurring threat to global banana production. Front Plant Sci 11:1–6

Larada JI, Pojas GJ, Ferrer LV (2018) Postharvest classification of banana (Musa acuminata) using tier-based machine learning. Postharvest Biol Technol 145:93–100. https://doi.org/10.1016/j.postharvbio.2018.06.004

Linting M, van der Kooij A (2012) Nonlinear principal components analysis with CATPCA: a tutorial. J Pers Assess 94(1):12–25. https://doi.org/10.1080/00223891.2011.627965

Liu A, Ghosh J, Martin C (2007) Generative oversampling for mining imbalanced datasets. DMIN 1:66–72

Lozano PZ, Bravo C, Ovalles F, Hernández RM, Moreno B, Piñango L, Villanueva JG (2004) Selección de un diseño de muestreo en parcelas experimentales a partir del estudio de la variabilidad espacial de los suelos. Bioagro 16(1):61–72

Lunardon N, Menardi G, Torelli N (2014) ROSE: a package for binary imbalanced learning. R J 6(1):82–92. https://doi.org/10.32614/RJ-2014-008

MacEwan RJ, Fitzpatrick RW (1996) The pedological context for assessment of soil quality. In: MacEwan RJ, Carter MR (eds), Soil Quality is in the hands of the land manager. Proceedings of an international symposium. Advances in soil quality for land management: science, practice and policy, 17–19 April 1996, University of Ballarat, Ballarat, Victoria, Australia, pp. 10–16

Manrique-Silupu J, Campos JC, Paiva E, Ipanaqué W (2021) Thrips incidence prediction in organic banana crop with machine learning. Heliyon 7(12):e08575. https://doi.org/10.1016/j.heliyon.2021.e08575

Martínez GE, Rey-Brina JC, Rodríguez D, Jiménez C, Rodríguez Y, Rumbos R, Pargas-Pichardo R, Martínez E (2020) Análisis de la situación fitopatológica actual de las musáceas comestibles en Venezuela. Agron Trop 70:1–20. https://doi.org/10.5281/zenodo.4323273

Mehlich A (1984) Mehlich 3 soil test extractant: a modification of Mehlich 2 extractant. Communications in Soil Science and Plant Analysis 15(12):1409–1416. v.1415 no.1412. https://doi.org/10.1080/00103628409367568

Meulman JJ, van der Kooij AJ, Duisters KLW (2019) ROS regression: integrating regularization with optimal scaling regression. Stat Sci 34:361–390. https://doi.org/10.1214/19-STS697

Miller A (2002) Subset selection in regression. Chapman & Hall/CRC, New York. https://doi.org/10.1201/9781420035933

Montenegro EJ, Pitti-Rodríguez JE, Olivares-Campos BO (2021) Identification of the main subsistence crops of Teribe: a case study based on multivariate techniques. Idesia (Arica) 39(3):83–94. https://doi.org/10.4067/S0718-34292021000300083

Munsell Color Company (1999) Munsell soil color charts. Munsell Color, Macbeth Division of Kollmorgen, Maryland

Nelson PE (1991) History of fusarium systematics. Phytopathology 81:1045–1048

Olivares B (2016) Descripción del manejo de suelos en sistemas de producción agrícola del sector Hamaca de Anzoátegui, Venezuela. Lgja 23(1):14–24. https://n9.cl/ycp08

Olivares BO (2018) Tropical rainfall conditions in rainfed agriculture in Carabobo, Venezuela. Lgr 27(1):86–102. https://doi.org/10.17163/lgr.n27.2018.07

Olivares B, Lobo D, Cortez A, Rodríguez MF, Rey JC (2017a) Socio-economic characteristics and methods of agricultural production of indigenous community Kashaama, Anzoategui, Venezuela. Rev Fac Agron (LUZ) 34(2):187–215. https://n9.cl/p2gc5

Olivares BO, Cortez A, Lobo D, Parra R, Rey J, Rodriguez M (2017b) Evaluation of agricultural vulnerability to drought weather in different locations of Venezuela. Rev Fac Agron (LUZ) 34:103–129

Olivares BO, Hernández R, Coelho R, Molina JC, Pereira Y (2018) Analysis of climate types: main strategies for sustainable decisions in agricultural areas of Carabobo, Venezuela. Sci Agropecu 9:359–369. https://doi.org/10.17268/sci.agropecu.2018.03.07

Olivares B, Pitti J, Montenegro E (2020a) Socioeconomic characterization of Bocas del Toro in Panama: an application of multi-variate techniques. Rev Bras Gest Desenvolv Reg 16(3):59–71. https://doi.org/10.54399/rbgdr.v16i3.5871

Olivares BO, Araya-Alman M, Acevedo-Opazo C, Rey JC, Cañete-Salinas P, Kurina FG, Balzarini M, Lobo D, Navas-Cortés JA, Landa BB, Gómez JA (2020b) Relationship between soil properties and banana productivity in the two main cultivation areas in Venezuela. J Soil Sci Plant Nutr 20:2512–2524. https://doi.org/10.1007/s42729-020-00317-8

Olivares B, Paredes F, Rey J, Lobo D, Galvis-Causil S (2021a) The relationship between the normalized difference vegetation index, rainfall, and potential evapotranspiration in a banana plantation of Venezuela. STJSSA 18:58–64. https://doi.org/10.20961/stjssa.v18i1.50379

Olivares B, Rey JC, Lobo D, Navas-Cortés JA, Gómez JA, Landa BB (2021b) Fusarium wilt of bananas: a review of agro-environmental factors in the venezuelan production system affecting its development. Agronomy 11(5):986. https://doi.org/10.3390/agronomy11050986

Olivares B, Vega A, Calderón MAR, Rey JC, Lobo D, Gómez JA, Landa BB (2022) Identification of soil properties associated with the incidence of Banana wilt using supervised methods. Plants 11(15):2070. https://doi.org/10.3390/plants11152070

Onweremadu EU, Akamigbo F, Igwe CA (2008) Soil quality morphological index about organic carbon content of soils in southwestern Nigeria. Trends Appl Sci Res 3(1):76–82

Paraskevaidi M, Cameron SJ, Whelan E, Bowden S, Tzafetas M, Mitra A, Kyrgiou M (2020) Laser-assisted rapid evaporative ionization mass spectrometry (LA-REIMS) as a metabolomics platform in cervical cancer screening. EBioMedicine 60:103017. https://doi.org/10.1016/j.ebiom.2020.103017

Pérez-Vicente L (2004) Marchitamiento por Fusarium (Mal de Panamá) en bananos: Una revisión actualizada del conocimiento presente sobre su agente causal. Fitosanidad 8:27–38

Phillips SJ, Anderson PR, Schapire RE (2006) Maximum entropy modeling of species geographic distributions. Ecol Model 190:231–259. https://doi.org/10.1016/j.ecolmodel.2005.03.026

Pitti J, Olivares BO, Montenegro E, Miller L, Ñango Y (2021) The role of agriculture in the Changuinola district: a case of applied economics in Panama. Trop Subtrop Agroecosyst 25:017. https://doi.org/10.56369/tsaes.3815

Pla I (1983) Metodología para la caracterización física con fines de diagnóstico de problemas de manejo y conservación de suelos en condiciones tropicales. Rev Fac Agron 32:1–91

Ploetz RC (1994) Panama Disease return of the first banana menace. Int J Pest Manag 40:326–336. https://doi.org/10.1080/09670879409371908

Ploetz RC (2006) Fusarium wilt of banana is caused by several pathogens referred to as *Fusarium oxysporum* f. sp. *cubense*. Phytopathology 96:653–656. https://doi.org/10.1094/phyto-96-0653

Ploetz RC (2015a) Fusarium wilt of banana. Phytopathology 105:1512–1521. https://doi.org/10.1094/phyto-04-15-0101-rvw

Ploetz RC (2015b) Management of fusarium wilt of banana: a review with special reference to tropical race 4. Crop Prot 73:7–15. https://doi.org/10.1016/j.cropro.2015.01.007

Ploetz RC, Pegg K (2000) Fusarium wilt in diseases of banana, abaca and enset. CABI Publishing, Wallingford, pp 143–159

Pocasangre LE, Ploetz RC, Molina AB, Vicente LP (2011) Raising awareness of the threat of fusarium wilt tropical race 4 in Latin America and the Caribbean. Acta Hort 897:331–337. https://doi.org/10.17660/ActaHortic.2011.897.45

Primpas I, Tsirtsis G, Karydis M, Kokkoris GD (2010) Principal component analysis: development of a multivariate index for assessing eutrophication according to the European water framework directive. Ecol Indic 10:178–183. https://doi.org/10.1016/j.ecolind.2009.04.007

Promusa (2022) Distribution of Foc TR4. http://wwwpromusaorg/Tropical+race+4+-+TR4#Distribution. Accessed 20 Feb 2021

Pulido-Moncada M, Helwig Penning L, Timm LC, Gabriels D, Cornelis WM (2014) Visual examinations and soil physical and hydraulic properties for assessing soil structural quality of soils with contrasting textures and land uses. Soil Tillage Res 140:20–28. https://doi.org/10.1016/j.still.2014.02.009

R Core Team (2020) R: a language and environment for statistical computing. R Foundation for Statistical Computing. Vienna, Austria.

Rey JC, Martínez G, Lobo D, Trejos J, Pocasangre L, Rosales F (2009) Aspectos sobre calidad y salud de suelos bananeros en Venezuela. Producción Agropecuaria 2:52–55

Rey JC, Martínez-Solórzano G, Ramírez H, Pargas-Pichardo R (2020) Relación de las condiciones agroecológicas de un lote de planicie lacustrina con la marchitez del banano Cavendish en Aragua, Venezuela. Agron Trop 70:1–12. https://doi.org/10.5281/zenodo.4346252

Rodríguez G, Núñez MC, Lobo D, Martínez G, Rey JC, Espinoza J, Muñoz N, González D, Rosales F, Pocasangre L, Delgado E (2006). Banana root health in lots with different productivity levels in a soil at the oriental coast of Maracaibo Lake, Venezuela. XVII Reunión Internacional ACORBAT: Banano un negocio sustentable. Joinville, Santa Catarina, Brasil. Nov 15–20. p. 355

Rosales FE, Pocasangre LE, Trejos J, Serrano E, Peña W (2008) Guía de Diagnóstico de la Calidad y Salud de Suelos. Bioversity International, Montpellier

Salzberg SL (1997) On comparing classifiers: Pitfalls to avoid and a recommended approach. Data Min Knowl Discov 1(3):317–328. https://doi.org/10.1023/A:1009752403260

Singh AP, Sahu P, Chug A, Singh D (2022) A systematic literature review of machine learning techniques deployed in agriculture: a case study of banana crop. IEEE Access 10:87333–87360. https://doi.org/10.1109/ACCESS.2022.3199926

Soil Survey Staff (2010) Keys to soil taxonomy, 11th edn. NRCS, Washington, DC

Soil Survey Staff (2017) Soil survey manual handbook 18. United States Department of Agriculture, Washington, D.C.

SQI-Soil Quality Institute (1996) Indicators for soil quality evaluation. USDA Natural Resources Conservation Service. Prepared by the National Soil Survey Center in cooperation with The Soil Quality Institute, NRCS, USDA, and the National Soil Tilth Laboratory, Agricultural Research

Stover RH (1962) Fusarium wilt (Panama disease) of banana and other Musa species, phytopathology paper 4. Commonwealth Mycological Institute, Kew Surrey, pp 17–29

Szymańska E, Saccenti E, Smilde AK, Westerhuis JA (2012) Double-check: validation of diagnostic statistics for PLS-DA models in metabolomics studies. Metabolomics 8:3–16. https://doi.org/10.1007/s11306-011-0330-3

Tibshirani R (1996) Regression shrinkage and selection via the Lasso. J Roy Statist Soc Ser B 58:267–288. https://n9.cl/pqv2k

Triba MN, Le Moyec L, Amathieu R, Goossens C, Bouchemal N, Nahon P, Rutledge DN, Savarin P (2015) PLS/OPLS models in metabolomics: the impact of permutation of dataset rows on the K-fold cross-validation quality parameters. Mol BioSyst 11:13–19. https://doi.org/10.1039/C4MB00414K

UNISDR (2015) Marco de Sendai para la Reducción del Riesgo de Desastres 2015–2030. 40 pp. UNISDR/GE/2015 – ICLUX ES 1a edición. https://n9.cl/q6kld

USDA (1995) Soil survey laboratory methods manual. USDA, Washington, DC

Van Eck NJ, Waltman L (2010) Software survey: VOSviewer, a computer program for bibliometric mapping. Scientometrics 84(2):523–538. https://doi.org/10.1007/s11192-009-0146-3

Van der Kooij AJ (2007) Prediction accuracy and stability of regression with optimal scaling transformations. Unpublished doctoral dissertation, Leiden University. Netherlands. https://n9.cl/ew8pz

van der Kooij AJ, Meulman JJ, Heiser WJ (2006) Local minima in categorical multiple regression. Comput Stat Data Anal 50:446–462. https://doi.org/10.1016/j.csda.2004.08.009

Vaquero MR (2005) Soil physical properties and banana root growth. In: Turner DW, Rosales FE (eds) Banana Root System: towards a better understanding for its productive management: proceedings of an international symposium. International Network for the Improvement of Banana and Plantain, Montpellier, pp 125–131

Weaver F, Angle J, Bottomley P (1994) Methods of soil analysis. Part 2. Microbiological and biochemical properties. Book series number 5. Soil Sci Soc Am, USA

Wissmann M, Shalabh D, Toutenburg H (2011) Role of categorical variables in multicollinearity in the linear regression model. J Appl Stat 19:99–115. https://n9.cl/r9d7u

Xia J, Wishart DS (2016) Using MetaboAnalyst 3.0 for comprehensive metabolomics data analysis. Curr Protoc Bioinformatics 55(1):14.10.1–14.10.91. https://doi.org/10.1002/cpbi.11

Xu QS, Liang YZ (2001) Monte Carlo cross validation. Chemom Intell Lab Syst 56(1):1–11. https://doi.org/10.1016/S0169-7439(00)00122-2

Xu J, Capretz LF, Ho D (2010) Building an OSS quality estimation model with CATREG. Inter J Comput Sci Eng 2:1952–1958. https://n9.cl/t3wlt

Yang J, Wang J, Qiao P, Zheng Y, Yang J, Chen T, Lei M, Wan X, Zhou X (2020) Identifying factors that influence soil heavy metals by using categorical regression analysis: a case study in Beijing, China. Front Environ Sci Eng 14(3):1–4. https://doi.org/10.1007/s11783-019-1216-2

Yang B, Zhang C, Cheng S, Li G, Griebel J, Neuhaus J (2021) Novel metabolic signatures of prostate cancer revealed by 1H-NMR metabolomics of urine. Diagnostics 11:149. https://doi.org/10.3390/diagnostics11020149

Chapter 3
Fusarium Wilt of Bananas: A Threat to the Banana Production Systems in Venezuela

Abstract The recent discovery of *Fusarium oxysporum* f. sp. *cubense* (*Foc*) Tropical Race 4 (TR4) (syn. *Fusarium odoratissimum*) in Colombia, Peru and Venezuela, as the main net exporting region of the world such as Latin America and the Caribbean, in addition to its permanent incidence in Asia, has greatly alarmed the banana export industry. The objective of this chapter is to synthesize the results of a reflective and practical review of the current state of knowledge concerning the main agro-environmental factors that may affect disease progression, and dissemination of this dangerous pathogen has been carried out, focusing on the Venezuelan Musaceae production systems. Environmental variables together with soil management and sustainable cultural practices are important factors affecting Fusarium wilt (FW) incidence and severity, excluding that the widespread dissemination of *Foc*, especially of its highly virulent TR4, is mainly caused by human activities. Additionally, risk analysis and climatic suitability maps for *Foc* TR4 in Venezuela have been developed. In all the cases reported in the literature, once a farm has been contaminated, disease management has proven to be extremely difficult and costly. This represents a particular threat to the livelihoods of small and medium-sized banana farmers in the affected countries, who often lack the financial means to implement measures in the face of yield loss alongside rising production costs. In this regard, prevention, rapid containment, and quarantine are essential for banana countries such as Venezuela.

Keywords Banana diseases · Climatic suitability · *Fusarium odoratissimum* · Pathogenic races · Epidemiology

3.1 Introduction

Bananas are of particular importance in some of the least developed, low-resource, food-deficit countries, where they contribute not only to household food security as a staple but also to income and employment generation as a crop commercial (FAO 2019). In the Venezuelan territory, there were 50,213 productive hectares of "Cavendish" bananas (Musa AAA) in 2020, with a production of 884,207 tons

B. O. Olivares Campos, *Banana Production in Venezuela*, The Latin American Studies Book Series, https://doi.org/10.1007/978-3-031-34475-6_3

destined for the local and export markets, whose average yield was 17.60 tons/ha (FAO 2019). The production of Musaceae in Venezuela is concentrated in four large areas: the western (Zulia, Mérida, Táchira, and Trujillo states), the southwestern (Barinas, Portuguesa, and Apure states), the central (Aragua, Carabobo, Yaracuy, Vargas, and Miranda states) and, to a limited extent, the eastern (Sucre, Delta Amacuro State) zones of the country (Olivares et al. 2020).

Banana production worldwide can be curtailed by several fungal diseases (e.g., Fusarium wilt or Panama disease, root rot) (Ploetz et al. 2021). Fusarium wilt (FW) of bananas (FWB) caused by *Fusarium oxysporum* f. sp. *cubense* (E.F. Sm.) W.C. Snyder and H.N. Hansen (*Foc*) is the main threat and limiting factor for different banana cultivars of economic and strategic importance all over the world (Kema et al. 2021).

Attention to FW in banana-producing countries has increased lately because it represents an important phytopathological problem of plantations in tropical and subtropical territories. *Foc*'s scientific literature has focused on biological aspects, the genetics of the pathogen, and advances in the genetic improvement of banana varieties (Kema et al. 2021). However, there is little scientific information on the biogeography and the relationships of the pathogen with climatic and edaphological factors; therefore, the information on these agro-environmental factors that affect the epidemiology of *Foc* are very limited in countries like Venezuela (Shabani et al. 2014).

Studies on epidemiological factors would represent an interesting contribution to the comprehensive understanding of the phytosanitary problem that *Foc* represents in bananas. Basically, the results of this chapter can be very useful for banana and disease management, obtaining an analysis of the relationship of *Foc* with other agronomic components. Therefore, this chapter aims to analyze the current state of knowledge of the main agri-environmental factors that affect the development and propagation of *Foc*, focusing on Venezuelan production. With the scientific knowledge compiled in this chapter, it will be possible to design certain sustainable management strategies to prevent or reduce the incidence of FW in banana plantations due to the recent declaration of *Foc* TR4 phytosanitary emergency in Venezuela.

3.2 Geographical Distribution: The Risk Posed by *Foc* Tropical Race 4 (TR4)

The disease was first described by Bancroft in 1876 in Australia and then by Ashby in 1913 in Costa Rica and Panama, where some 80,000 ha of the cultivar Gros Michel (AAA) were devastated in Latin America and the Caribbean by *Foc* R1 (Pérez-Vicente 2004). The origin of *Foc* was in Southeast Asia, and it has co-evolved with Musaceae in its center of origin, being reported in all banana-producing regions (Fig. 3.1), except for the South Pacific Islands, Somalia, and coastal countries from the Mediterranean Basin (Bentley et al. 1998; Maryani et al. 2009).

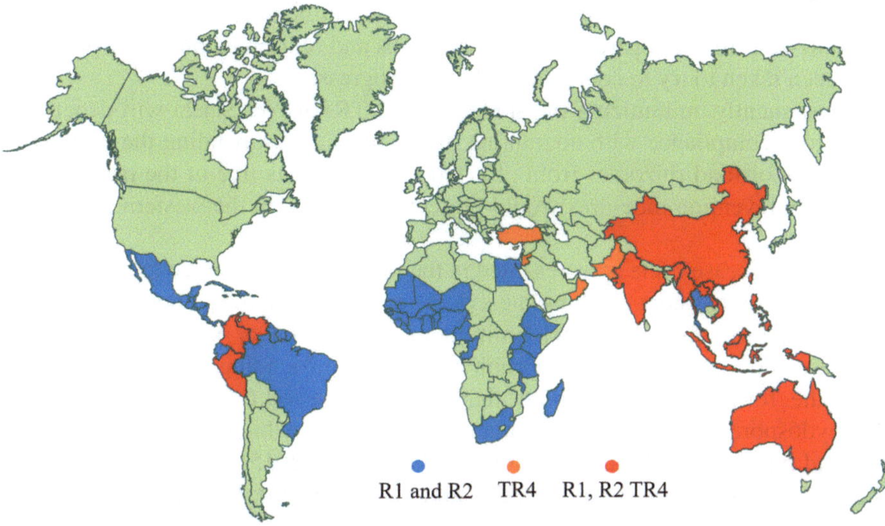

Fig. 3.1 Geographical distribution of *Fusarium oxysporum* f. sp. *cubense* (*Foc*) races. (Source: Adapted from CABI/EPPO (2015), EPPO (2014), and Promusa (2021))

Foc has three races, R4 being the most virulent, which is further subdivided into tropical (TR4) and subtropical (SR4) races. *Foc* SR4 attacks banana cultivars of the "Cavendish" group in subtropical regions such as Taiwan, the Canary Islands (Spain), South Africa, and Australia. The studies developed by Buddenhagen (2009), Ploetz (2006), and Dita et al. (2017) show that TR4 is a different *Foc* pathotype, highly virulent even under unfavorable environmental conditions for disease incidence. On the other hand, *Foc* TR4 attacks banana cultivars of the "Cavendish" group in Australia and tropical regions of Southeast Asia, such as China, the Philippines, Malaysia, and Indonesia (Fig. 3.1) (Ploetz and Pegg 2000; Molina et al. 2009; Garcia-Bastidas et al. 2014). Indeed, very recently, this *Foc* TR4 was found to be genetically distant from the other races and has been described as *Fusarium odoratissimum* (Maryani et al. 2018).

Today, *Foc* R4 is causing unexpected damage to the cv. Cavendish in the tropics severely impacting production and exports in Southeast Asia and the "Cavendish" market in the Western Hemisphere (Kema et al. 2021; Maryani et al. 2009). Given this scenario, the production of banana-producing countries in Latin America and West Africa will be affected. In recent years, the recent appearance of *Foc* TR4 in the municipalities of Dibulla and Riohacha (Guajira) in Colombia (Fontagro 2021) and in the Department of Piura (Peru) put at risk the production and export of bananas from Colombia and Peru but also for other producing countries in the region such as Ecuador and Venezuela.

In January 2023, the presence of *Foc* TR4 in Venezuela was officially declared for the first time according to the declaration of a state of phytosanitary emergency in-formation provided by INSAI (INSAI official written communication, 19 January 2023; Martinez et al. 2023). The sources of infection were found in the states of Aragua, Carabobo, and Cojedes, which is indicative of the advance of the pathogen in the American continent. However, no further scientific information has been published to date. In this regard, pertinent measures and reinforced biosafety protocols have been taken to try to contain or delay its progress.

Consequently, minimizing the spread of *Foc* TR4 in Venezuela will depend on mandatory compliance with quarantine measures, such as avoiding the transfer of banana shoots and rhizomes from affected areas. to areas free of the pathogen. In this way, prevention measures would protect production in the western part of the country.

The *Foc* disease cycle is represented in the first place by the infection of the plant through the root system reaching the xylem vessels, where the pathogen grows and multiplies, occluding them and limiting the absorption of nutrients and water. Likewise, Foc can also grow saprophytically in surrounding tissues as diseased plants die, forming chlamydospores that remain in the soil for a long time. Second, chlamydospores are stimulated by plant root exudates, which subsequently germinate and infect nearby healthy plant roots (Ghag et al. 2015). Third, after germination and root infection, more mycelia and chlamydospores are produced, which infect the secondary or tertiary roots of the banana.

In susceptible cultivars, the infection becomes systemic through the vascular system of the corm, pseudostem, and stem because the fungus is not blocked by defense mechanisms (Ploetz and Pegg 2000; Ghag et al. 2015). Young plants have been shown to be the most susceptible (Ploetz 1994) due to the presence of younger roots susceptible to *Foc* infection (Dita et al. 2018).

The main symptom of *Foc* is the yellowing and wilting of the leaves due to the blockage of the flow of water in the xylem vessels. Distinctive symptoms appear within the pseudostem, the main features of which are ring-shaped lines of brown, red, or yellow. Brown streaks or specks appear on the corm. Infected plants usually do not produce fruits, or the size of the fruits decreases considerably (Ploetz and Pegg 2000) (Fig. 3.2).

Foc being a soilborne pathogen, movement of infested soil can disperse soil particles infested by fungal structures (e.g., spores and mycelium). In addition, irrigation water, the weevil *Cosmopolites sordidus*, and other secondary hosts such as ornamental plants and weeds can contribute to the spread of the pathogen (Kema et al. 2021; Waite and Dunlap 1953). The movement of plants as planting material also plays a significant role in pathogen dissemination (Ploetz et al. 1994). A summary of the main *Foc* dispersal means is discussed below (see Sect. 3.4.6).

3.3 A Bibliometric Analysis of the Scientific Production of *Foc* TR4 in Latin America and the Caribbean (LAC)

A total of 284 publications on the topic of *Foc* TR4 were identified in the Web of Science (WOS) database between 2019 and 2022, including 243 (85.6%) original research articles, 11 (3.9%) review articles, 22 (7.7%) conference papers, and eight

(a) (b) (c) (d)

Fig. 3.2 *Fusarium oxysporum* f. sp. *cubense* symptoms in plants of banana clone "Manzano" growing at "El Diamante" Farm, El Cenizo Irrigation System, Sabana Mendoza, Trujillo State of Venezuela. (**a, c, d**) Internal brown to red brick discoloration of the vascular system. (**b**) The first external symptoms of Fusarium wilt are chlorosis and the death of the oldest leaves, which usually bend and collapse against the pseudostem. (Pictures by Gustavo Martínez) (Trujillo-Venezuela, 2009)

contributions from other forms of publications including notes, book chapters, and erratum. Of all the published articles, only those that mentioned the LAC problem or according to the origin of the authors in the WOS central database were selected, obtaining a total of 92 documents, with contributions from Brazil (22.8%), Colombia (13.0%), and Mexico (7.6%) having the highest proportions, followed by the scientific production of Costa Rica, Cuba, Ecuador, Peru, Uruguay, and Venezuela.

The bibliometric analysis of the keywords is shown in Fig. 3.3a. Keywords provided by the article authors and occurring more than five times in the WOS core database were enrolled in the final analysis. Of the 1074 keywords, 37 reached the threshold. The keywords that appeared the most were "Fusarium" (total bond strength 119) and "Musa" (total bond strength 103), which had a strong link with "microbiology" and "plant disease" (Fig. 3.3a). Four clusters were identified; the first cluster has seven items, the keyword "non-human" (link 19 and total strength of link 78) being the one with the greatest influence and occurrence (17 times), followed by "Fusarium wilt" (link 20, total bond strength 59, and occurrence 33 times). The second cluster has seven items; "plant disease" (link 19, total strength of link 81, and occurrence 14 times), "Musa" (link 18, total link strength 103, and occurrence 21 times), and "microbiology" (link 18, total link strength 80, and occurrence 12 times) have the greatest influence on the network. The third cluster has five items, with "fungi" being the most influential term (link 17, total link strength 57, and occurrence 22 times). Finally, the fourth cluster has only four items, with "pathogenicity" being the most influential term (link 16, the total link strength 47, and occurrence eight times) and "cultivar" (link 16, total strength of link 46, and occurrence seven times).

The bibliometric analysis of topics and trending topics is indicated in Fig. 3.3b; three subjects of *Foc* TR4 studies were found. Group 1 involved publications describing the diagnosis of *Foc* TR4 and the characteristics of the cultivar. Group 2 involved publications in the genetic field and pathogenesis. Group 3 involved those contributions that investigated the effects on growth and control. Figure 3.3b shows the network map of trending topics based on keywords used from January 2019 to September 2022. The indicator shows current posts from purple to yellow. More studies focusing on the effects on growth and biological control, for example, have recently been published.

The threat of *Foc* TR4 remains latent and at risk of spreading throughout LAC until now, and the new cases of *Foc* TR4 reported in Colombia and Peru make the situation very worrying. To fight this disease, academia in LAC with international allies must join this "battlefield" as soon as possible to provide recommendations and suggestions to prevent the appearance of the disease in Musaceae.

The contributions in LAC demonstrate a proactive research community in search of the effective control of *Foc* TR4. Together, the documents provide an overview of the current understanding of the biology and pathogenesis of TR4, its management, and control alternatives; in this sense, integrated and innovative solutions are required to be adopted by all interested parties to build the sustainability of Musaceae systems in the future. There are some topics in which further development is expected, such as the use of new resistant genotypes (Dita et al. 2018); identification of new virulent strains (Martínez et al. 2020); the environmental suitability of the disease (Olivares et al. 2021); and the diversification of the genetic bases of the banana industry combined with other types of solutions, such as soil management or the integration of bananas with other crops (Kema et al. 2021).

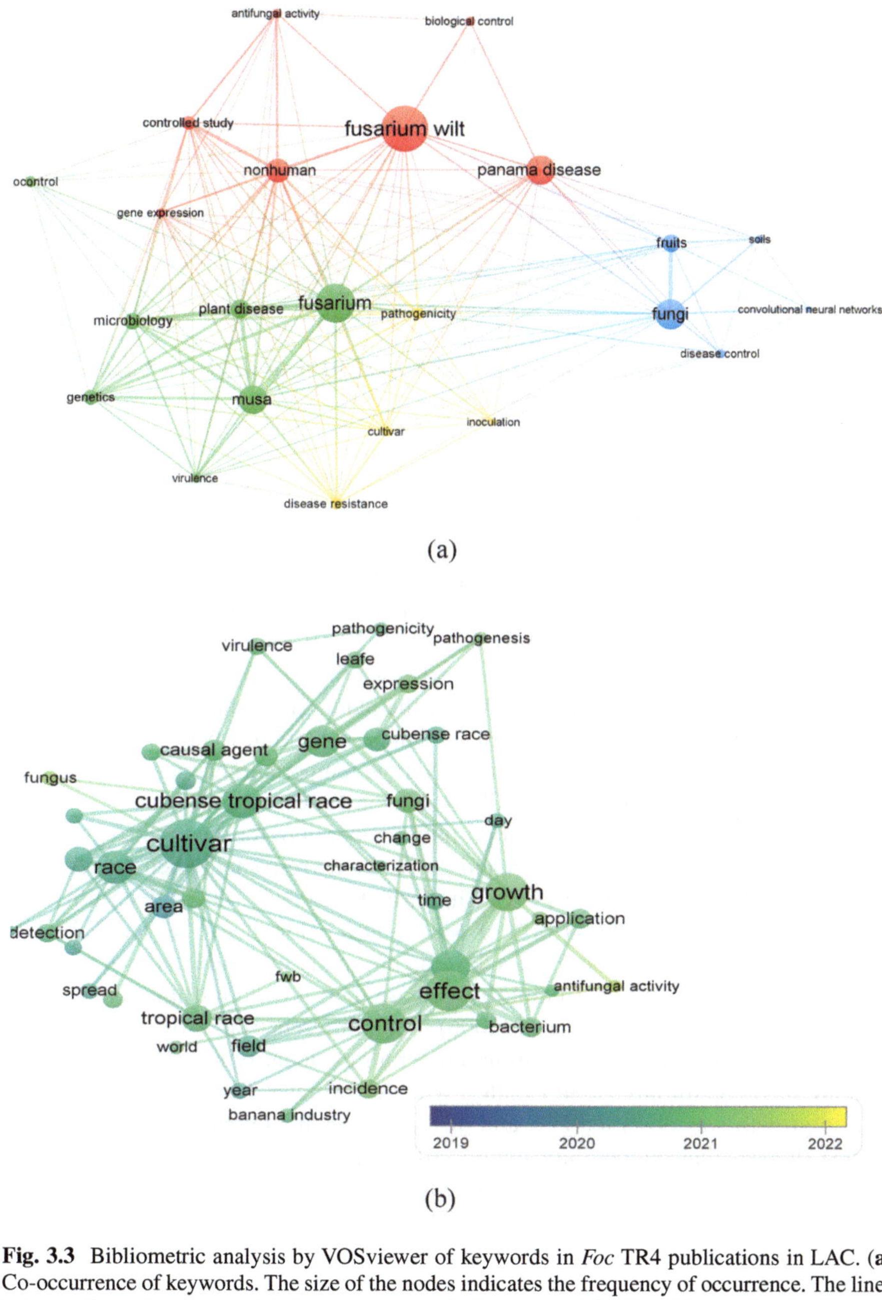

Fig. 3.3 Bibliometric analysis by VOSviewer of keywords in *Foc* TR4 publications in LAC. (**a**) Co-occurrence of keywords. The size of the nodes indicates the frequency of occurrence. The lines between the nodes represent their co-occurrence in the same publication. The smaller the distance between two nodes, the higher the number of co-occurrences of the two keywords. (**b**) Network map of trending topics according to the keywords used from January 2019 to September 2022. The size of the circles represents the frequency of occurrence of the keywords. The distance between the two circles indicates their correlation

3.4 Factors That Affect the Development of Fusarium Wilt Epidemics in Banana Crops

Several abiotic and biotic factors of physical, chemical, microbial, climatic, or even sociocultural nature have been shown to have an important role in the interactions between the banana crop, the pathogen, and the soil environment. As a result of those complex and site-specific interactions, the FW of bananas may develop or not, leading to detrimental levels for banana production.

3.4.1 *Environmental Conditions: Current and Future Climate Change Scenarios*

In general, banana growth is not significantly limited by solar radiation or temperature; precipitation is the most important climatic factor for rainfed plantations (Olivares 2018). However, these factors can directly influence the occurrence of certain tropical diseases such as Black Sigatoka (*Mycosphaerella fijiensis* Morelet), Yellow Sigatoka (*Mycosphaerella musicola* Leach et Mulder), and False Panama disorder (of unknown etiology) (Martínez et al. 2020). But in the case of banana-producing farms located in the extratropical zone, the temperature would represent a limiting factor for the adequate growth of the plants (Bosman 2016). The banana cultivation is characteristic of tropical lowlands, in latitudes below 10°, altitudes around 100 m above sea level, minimum average temperature values of 19 °C, and total rainfall exceeding 100 mm/month (Robinson 1996).

According to Nelson (1991), the optimum temperature for the in vitro growth of *F. oxysporum* isolates is between 25 and 28 °C, being restricted when the temperature exceeds 33 °C or is below 17 °C. Similarly, Pérez et al. (2003) determined the effect of temperature on the growth of *Foc* isolates belonging to R1 and R2 races. They found that isolates of both races develop optimally in a wide range of temperatures between 23 and 29 °C, to rapidly decrease their growth outside this range. In a different study, Groenewald et al. (2006) found differences in growth rate in vitro of *Foc* isolates at different incubation temperatures, identifying an optimum temperature of 25 °C for almost all isolates, while growth was limited at 10 and 35 °C and no growth was observed at 5 and 40 °C. Under field conditions, Brake et al. (1995) showed that temperature mainly affects plant growth rather than directly affecting the virulence of the pathogen. However, under temperature values around its optimal values for growth, higher severity of the disease would be expected since the growth and reproduction of the fungus would be favored by these environmental conditions. Moreover, high-temperature values would also favor water stress in the plants, enhancing the severity of *Foc* symptoms. Ploetz (1990) showed that temperature had an important effect on the development of FW disease. In this context, the level of susceptibility/resistance to *Foc* R1 in certain banana cultivars, such as "Lacatan" or "Gros Michel," would largely depend on temperature. That is, in certain banana cultivars, the level of disease would increase under a temperature range optimum for disease development compared to that developed when suboptimal temperatures prevail.

Current research suggests that one of the main consequences of climate change would be the increase of temperatures and the alteration of rainfall patterns that could modify the incidence of pests and diseases and consequently impact the productivity of crops. Although the information on the occurrence records available for the different *formae speciales* of *F. oxysporum* for *Foc* in banana crops is limited, some studies have modeled the future distribution of other *formae speciales* of *F. oxysporum* estimating that at the species level, this fungus would find suitable areas for growth in North Africa and Middle Eastern and European countries for the years 2050 and 2100 (Shabani et al. 2014). In a similar analysis, Pérez-Vicente and Porras (2015) estimated different climate change scenarios in Cuba, leading to a reduction in the geographic extent of the *Foc* pathogen, although FW would increase its severity when present.

Projected climate change in Venezuela is expected to increase the occurrence of potential areas for Foc TR4. The decrease in rainfall during the driest season and the wettest month due to climate change could favor the occurrence of *Foc* TR4 in a banana-producing area, as well as an increase in rainfall during the warmest quarter (Fig. 3.4a). The decrease in rainfall during the driest and wettest quarter and the increase in maximum temperatures in the national territory could cause a water stress condition that makes the plant highly susceptible to the disease (Fig. 3.5a).

In Venezuela, a set of Musaceae (banana and plantain) production systems have been developed that respond to the climate and soil conditions of the different product areas. Most of them are highly dependent on the rainy season, characterized by the predominance of monoculture with little genetic diversity and generally poor management practices. Problems of soil and water degradation or contamination are attributed to these practices, in combination with the direct effects of climate variability and climate change (MARN 2005; Ovalles et al. 2005; Olivares et al. 2018).

According to the opinion of climate change experts, an increase in average temperature, a decrease in total precipitation, and the number of wet months in various areas of the country are generally expected, as well as a more frequent occurrence of extreme events (Ovalles et al. 2008). What has been described above has a direct impact on agriculture, with the potential for direct damage, especially in bananas: reduction in yields, affectation of cycles of pests and diseases of agricultural interest, and major changes in the spatiotemporal distribution of agricultural systems.

The area harvested for bananas (4481 ha) for the Central Region, concentrated mostly in the state of Aragua (Olivares et al. 2017a), will be directly and indirectly affected by a change in the magnitude of the rains; that is, those areas where there was rainfall of 1600–2000 mm/year will disappear and a lower range of 800–1200 mm/year will dominate according to the UKTR Model (Martelo 2003; Ovalles et al. 2008). The water deficit will increase from one million ha to 4.7 million ha, whose drought events will have repercussions on banana yields (Olivares and Zingaretti 2018, 2019). This situation will mean that irrigated areas can

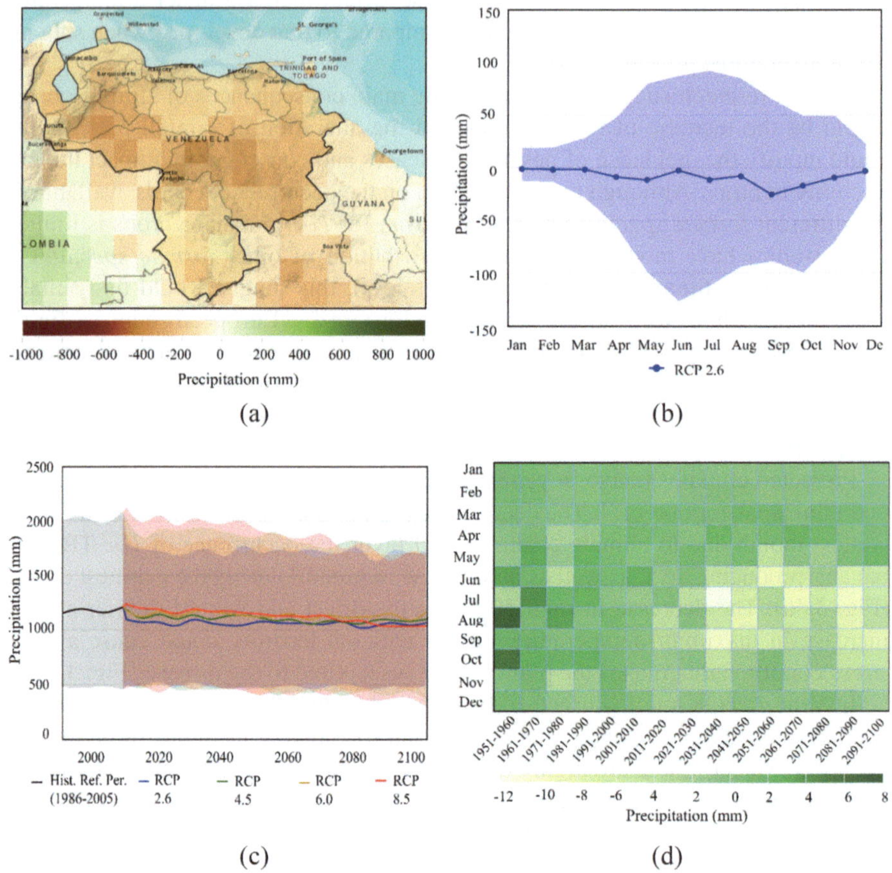

Fig. 3.4 (**a**) Spatial distribution of projected precipitation anomaly for 2020–2039 (annual) in Venezuela (reference period: 1986–2005), RCP 2.6, multi-model ensemble according to the data source from Taylor et al. (2012); (**b**) projected precipitation anomaly for 2020–2039 (annual) in Venezuela; (**c**) projected precipitation anomaly in Venezuela (reference period: 1986–2005), RCP 2.6, RCP 4.5, RCP 6.0, RCP 8.5, multi-model ensemble; (**d**) projected precipitation anomaly in Venezuela (reference period: 1986–2005), RCP 2.6, multi-model ensemble

compete with tourism and other uses of water, requiring the need to change irrigation systems for more efficient ones in some large agricultural production units (Viloria et al. 2023; Cortez et al. 2018; Olivares et al. 2017b). However, there is uncertainty in the predictions of the models (especially precipitation) so they should be interpreted as a probable trend whose absolute values still need to be refined.

From the models to understand the feedback of the Earth's carbon cycle in response to anthropogenic emissions, which cause changes in the atmospheric concentrations of greenhouse gases and aerosols, the Representative Concentration Pathways (RCP) arise, presented in the Coupled Model Intercomparison Project 5 (CMIP5), describing four different paths of the twenty first century (Van Vuuren

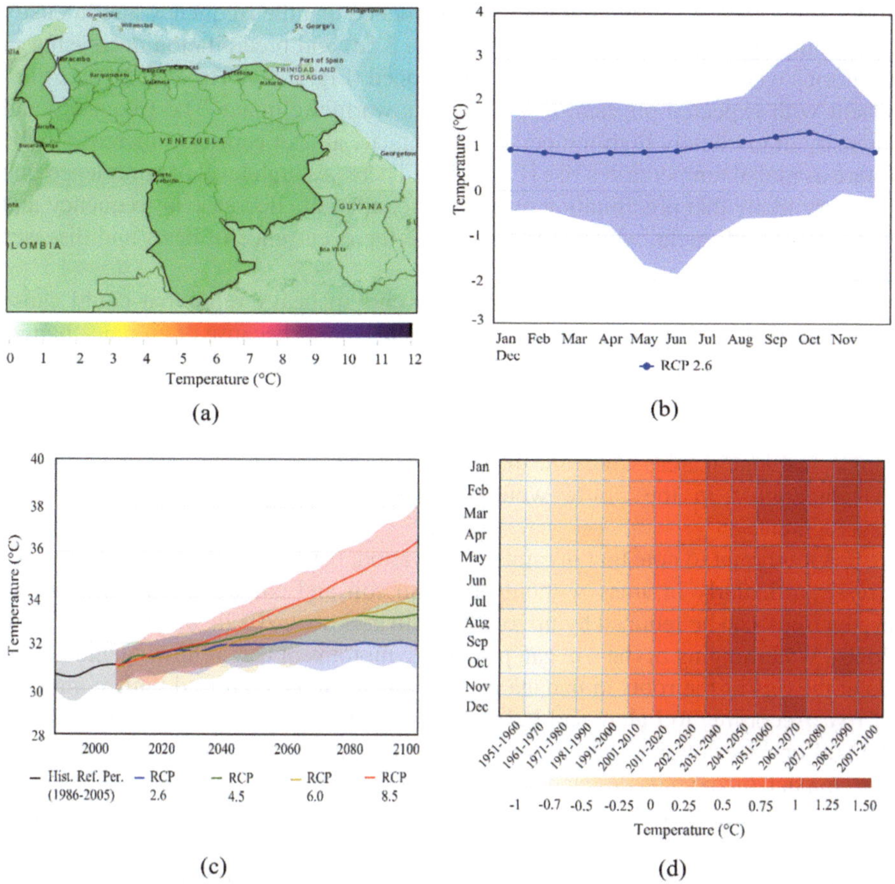

Fig. 3.5 (**a**) Spatial distribution of projected maximum temperature anomaly for 2020–2039 (annual) in Venezuela (reference period: 1986–2005), RCP 2.6, multi-model ensemble according to the data source from Taylor et al. (2012); (**b**) projected maximum temperature anomaly for 2020–2039 (annual) in Venezuela; (**c**) projected maximum temperature in Venezuela (reference period: 1986–2005), RCP 2.6, RCP 4.5, RCP 6.0, RCP 8.5, multi-model ensemble; (**d**) projected maximum temperature anomaly in Venezuela (reference period: 1986–2005), RCP 2.6, multi-model ensemble

et al. 2011). The RCPs include a strict mitigation scenario (RCP 2.6), two intermediate scenarios (RCP 4.5 and RCP 6.0), and a scenario with high greenhouse gas (GHG) and aerosol emissions (RCP 8.5). Scenarios without additional efforts to limit emissions ("baseline scenarios") lead to paths ranging between RCP 6.0 and RCP 8.5. Each RCP shows the planet trapping progressively higher amounts of energy from RCP 2.6 (lowest) to RCP 8.5 (highest) (WCRP 2021).

The impacts related to climate change are evident in all regions and in many important sectors of Venezuela such as agriculture and food security; therefore, the impacts are expected to be increasingly disruptive in the coming decades (Figs. 3.4b and 3.5b). There is a very high level of confidence that the frequency and intensity of extreme heat episodes are increasing (Fig. 3.5c) and precipitation decreasing in territories where rainfed agriculture is developed (Fig. 3.4c). These trends are consistent with expected physical responses to a warmer climate. The frequency and intensity of extremely high-temperature events are almost certain to increase in the future as global temperatures rise (Fig. 3.5d). There is a high level of confidence that it is very likely that precipitation events will continue to decrease in frequency and intensity in Venezuela, which could facilitate the occurrence of pests and diseases (Fig. 3.4d).

Additionally, in conditions of extreme events of heavy rains that could cause flooding, whose incidence might increase in some areas because of climate change, *Foc* could severely affect areas with susceptible cultivars such as the Burro (Bluggoe, ABB) types. Shabani et al. (2014) and Pérez-Vicente and Porras (2015) also indicate that the reported temperature range for the pathogen exceeds the temperature range prevalent in areas where the disease occurs in Cuba, and the expected increase in temperature in the future would render these areas not suitable for FW development.

It is important to note that in rainfed cultivation, banana is very sensitive to water stress, particularly in areas without optimum soil and climate conditions, whose productivity can be reduced by up to 50% (Olivares et al. 2017a, b; Nansamba et al. 2020). Waterlogging of the soil can produce symptoms of leaf yellowing and necrosis at the tips of the roots in banana plants, which can be confused with symptoms of FW. According to Lahav and Israeli (2000), conditions of excessive moisture in the soil or waterlogging would predispose the plant to infection by the pathogen. The impact of oxygen deficiency on the interaction with *Foc* was evaluated by Aguilar et al. (2000) based on changes in the activities of enzymes involved in phenol metabolism (phenylalanine ammonia lyase (PAL) and peroxidase (PER)) in banana cultivars differing in their reaction to *Foc*. Infected plants were subjected to hypoxia-induced changes in peroxidase activity, which correlated with their resistance to FW. However, a breakdown of resistance to *Foc* of cv. Williams occurred when the soil was waterlogged. On the contrary, it has also been documented that flooding longer than 18 months destroys *Foc*'s reproductive structures (Ploetz 1990). It is well established that *Fusarium* spp. can survive in the soil for long periods when unfavorable drought conditions occur (Summerell et al. 2011). However, in arid or semi-arid conditions, with rainfall below 500 mm/year, it has been shown that soil moisture levels and temperature conditions can be unfavorable for *Foc* growth and development (Nansamba et al. 2020). Although mycelium and conidia survive for a short time in very dry soils, this stress situation represents a starting point for the development of *Foc* chlamydospores that can remain dormant for approximately 20–30 years (Brake et al. 1995).

3.4.2 Land and Soil Physical Properties

Figure 3.6 shows the correlations between categorical dimensions representing them as flows (agri-environmental factor, variable, and property) visually linking categories with shared elements. Each rectangle represents a single value in the selected dimension; its height is proportional to its value. It is observed that the studies related to the properties of the soil are the ones with the highest proportion, followed by climatic studies and those of agronomic management.

Bosman (2016) studied the relationship between the slope in the field and the incidence of *Foc* in banana plantations in tropical soils of Costa Rica, finding that steep slopes have more erosion and nutrient losses than gentle slopes, where sediments often accumulate. Erosion also contributes to the spread of *Foc* inoculum that is transported with sediments and runoff. In addition, steep slope and slope location influence the water availability in the soil and the amount of runoff water, and the movement of *Foc* inoculum close to the root system (Rishbeth 1957).

Soil structure, often expressed as the degree of stability of aggregates, exerts important influences on the edaphic conditions and the environment and results from the rearrangement, flocculation, and cementation of soil particles. It is mediated by soil organic carbon (SOC), biota, ionic bridging, clay, and carbonates (Bronick and Lal 2005). Li et al. (2018) analyzed two typical banana-growing soils (ultisol and inceptisol), which were either suppressive or conducive to the FW of bananas from Hainan, China. They found that the suppressive soils had significantly more >2 and < 0.053 mm aggregates, had a comparatively even size distribution of

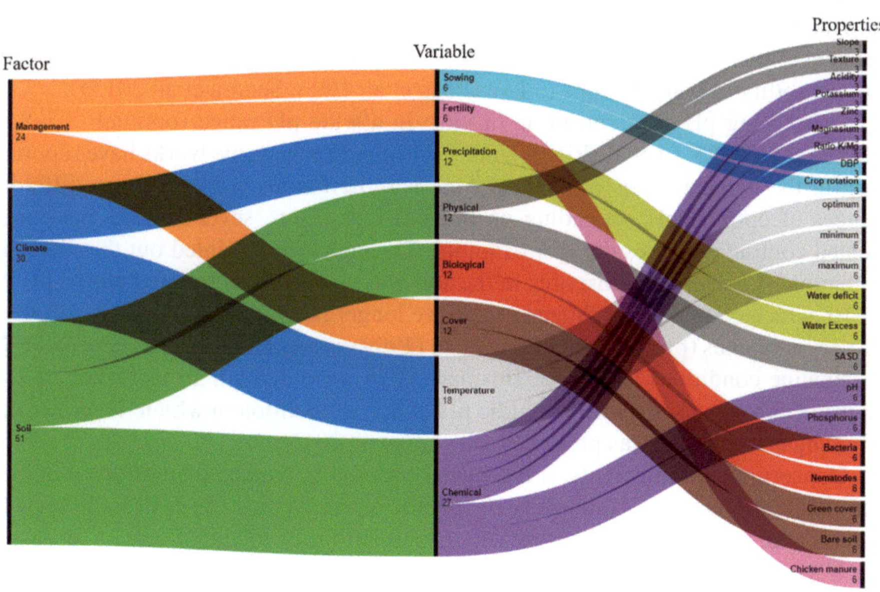

Fig. 3.6 Alluvial diagram of research on agro-environmental factors and *Foc* TR4 ($n = 105$)

aggregates within the range of 0–0.25 mm, and had higher total carbon, total nitrogen, and soil enzyme activity in the aggregates.

In the Canary Islands, Spain, Domínguez et al. (2001) found that soils suppressive to *Foc* had high electric conductivity (EC), higher levels of clay, and soluble Na. In the same region, on volcanic soils, Domínguez et al. (2008) found a clear separation between areas with and without FW, finding that the soluble K/Na ratio was always greater in affected areas, which is correlated with higher amounts of clay-sized particles and the increase of water-stable aggregate mass in these diseased areas. Moreover, the low potential buffering capacity for K observed in diseased areas suggests that massive K fertilization might exert a negative effect on disease development in banana plants. In Brazil, Deltour et al. (2017) showed that soils with a higher level of suppressiveness to *Foc* R1 are characterized by higher clay content and higher pH, which suggests that soils with heavy texture could be less prone to the development of FW than sandy soils with lower pH.

3.4.3 Soil Chemical Properties

Multiple soil chemical properties including pH, organic matter content, and the availability of some micronutrients in the soil are related to the suppressiveness of FW in certain soils. However, this effect largely depends on the soil type and climate, which makes it not possible to generalize the effect of those soil properties on the development of FW diseases. Nevertheless, for the specific case of FW of bananas, some research works have identified some specific chemical soil properties including pH, organic matter, potassium, phosphorus, nitrogen, and magnesium content (either by excess or deficient levels) as key factors contributing to reduce the susceptibility of banana crops to FW (Pegg et al. 1996; Segura et al. 2015a).

Concerning pH, *Foc* can grow in vitro in a wide pH range (optimal, 7.5–8.5) (Brake et al. 1995; Li et al. 2018). Under field conditions, some works have reported that in the banana zones of Central America, the fungus seems to have different ranges of optimal pH depending on the soil type. Thus, several works have presented contradictory results. For example, Chuang (1991) pointed out that the germination of *Foc* chlamydospores in the soil was negatively correlated with a pH of 8–10, which caused the pathogen to survive longer in alkaline soils (pH 8–10) than in very acid ones (pH 2–4), and Peng et al. (1999) showed a higher incidence of FW in alkaline conditions. However, Bosman (2016) recorded a high incidence of *Foc* in tropical banana soils with an acidic pH of 5.1, a condition in which the growth of the fungus in the soil was promoted (Segura et al. 2015b).

Concerning soil nutrients, it has been observed in banana plantations that certain fluctuations in the availability of specific nutrients may cause stress in banana plants and, consequently, increase their susceptibility to the attack of diseases, such as FW (Gutiérrez et al. 1983). Among the macronutrients, potassium has been shown to have a direct and indirect influence on the incidence of FW. Soils with potassium deficiencies are directly correlated with a high incidence of the disease (Huber et al. 2012). According to Domínguez et al. (2008), the content of potassium in soil solution together with the presence of fine clay particles in the soil induces the suppression of FW in banana plantations in volcanic soils grown in arid and semi-arid regions, such as the Canary Islands.

The areas with a high incidence of *Foc* in Costa Rica, have a lower average phosphorus concentration (20 mg/L) than the areas with low incidence, resulting from this a nutrient that is very important not only for the health status of the plants but also for the incidence of FW (Segura et al. 2015a). In the Canary Islands, the study conducted by Borges et al. (1991) showed that zinc application to the soil significantly reduced the incidence and severity of FW in bananas in that area.

Concerning magnesium, a higher incidence of diseases in banana fields is related to high concentrations of this element in the soil. Therefore, in soil, a low level of magnesium might be recommended to increase plant resistance to FW. Despite the above, magnesium is usually applied at high concentrations as fertilizer to increase banana growth.

Interestingly, rather than the concentration of specific nutrients in the soil, some authors highlighted the importance of the potassium/magnesium ratio in the reaction of banana crops to *Foc*. Thus, Borges et al. (1983) reported that banana soils infected by *Foc* in the Canary Islands have the highest values of the potassium/magnesium ratio (0.67) compared to soils with healthy plants (0.48). The absolute values of potassium in these soils are commonly high due to the frequent applications of potassium salts as fertilizers. On the other hand, the absolute values of magnesium were low in soils that presented the disease. Other studies have indicated that the potassium/magnesium ratio ranges between 0.55 and 0.81 and is associated with serious losses due to FW in banana plantations in the Canary Islands (Gutiérrez et al. 1983).

3.4.4 Crop Management

At the banana production unit level, agroecosystem management seems to have a major role in the incidence of FW. For example, the use or establishment of a green cover in banana plantations based on fodder peanut (*Arachis pintoi*), carpet grass (*Axonopus affinis*), and indigenous native grass (*Paspalum conjugatum*) has been shown to have a positive impact on reducing the incidence of FW (Rasiah et al. 2009; Pattison et al. 2014). In line with this observation, it is commonly observed that a high percentage of bare soil (30.1%), which is related to a higher incidence of FW in the banana fields of Central America (Bosman 2016). Experiences in this region and in Australia suggest that the presence of bare soil should be avoided and replaced by a green or brown cover (Pattison et al. 2014).

Practices like biological disinfestation of the soil based on the incorporation of the cover crop or organic amendments (rice straw) of easy decomposition, flooding with irrigation, and covering with a plastic film lead to anaerobic conditions, which helps to control the attack of pathogens including *Foc* (Huang et al. 2015).

Intercropping and crop rotation are old practices that are used for disease control (Olivares et al. 2017a; b). For the FW of banana, the rotation with Chinese chives (*Allium tuberosum*) is a cultural practice capable of reducing the incidence of *Foc* disease in China, significantly inhibiting the growth and causing the death of *Foc* spores (Zhang et al. 2013).

Another aspect to consider is plant density. According to Bosman (2016), if the plantation has a higher plant density (<3 m of plant spacing), the chances of the plants becoming infected are higher. A smaller distance between banana plants generates negative impacts on their health through competition and increases the chances of *Foc* infection. However, the opposite effect may occur due to lower microbial activity, since the root activity is reduced and movement of water in the soil limited, resulting in a condition of less fertile soils and, therefore, more susceptibility to *Foc* attack (Dita et al. 2018).

Finally, there are studies indicating that the addition of chicken manure increases the inoculum of *Foc* in soil and the incidence of FW, despite being an agricultural practice that improves the soil fertility in banana plantations infected by *Foc* (Nasira et al. 2003).

3.4.5 Soil Biota

The proportional size of abundance of microbial populations in soils appears to have a large influence of FW on bananas (Shen et al. 2015a; Shen et al. 2018). On this premise, microbial soil populations are essential to suppress *Foc* in this environment, while on the other hand, the physical and chemical properties of the soil affect the growth and development of the microbial population in the plant rhizosphere.

Bacterial and fungal communities were mainly determined by the organic matter content in banana soils in China (Shen et al. 2015a). These bacterial and fungal communities were significantly altered after long-term banana monoculture, indicating that the increase in fungal richness showed a significant correlation with the high incidence of FW disease. In this regard, Deng et al. (2013) showed that the metabolic characteristics of the microbial communities present in a banana plantation in China were significantly different when comparing healthy and diseased plants on the same banana plot.

Nowadays, the use of next-generation sequencing techniques is facilitating the study and the understanding of the plant-associated microbial communities and their shifts under varying conditions. Furthermore, the plant-associated microorganisms or microbiome is recognized as a key factor behind the health of the plants (Pirttila 2016). Although there is still a lack of knowledge concerning the relationships between the microbiome profiles of the banana plant and the rhizosphere environment, some recent works are providing insights into the effects of banana-associated microbiome and the development of FW. Thus, some studies

(Pirttila 2016; Koberl et al. 2015) found a determining beneficial role of Gammaproteobacteria present in banana soils in Central America, concluding that some members of this bacterial class were associated with the lack of successful infection of banana plants by *Foc* in soils infected by this pathogen. Additionally, Koberl et al. (2017) found an increase in the *Pseudomonas* and *Stenotrophomonas* populations in healthy banana plants, whereas FW-diseased plants showed an increase in Enterobacteriaceae. More recent studies have found that the phylum Acidobacteria was significantly elevated, but Bacteroidetes was significantly reduced in banana soils suppressive to FW. Additionally, certain bacteria belonging to the genera *Gp4*, *Gp5*, *Chthonomonas*, *Pseudomonas*, and *Tumebacillus* were specifically enriched in suppressive soils, whereas *Gp2* was reduced (Shen et al. 2015b). However, it is important to point out that the exact mechanisms responsible for FW microbial suppression are probably due to complex microbial communities more than to specific bacterial genera.

The application of biological control agents (BCAs) not only effectively controls soilborne pathogens such as *Foc* but also significantly promotes plant growth and increases plant biomass. For the specific case of FW of bananas, several studies have dealt with the application of BCAs at the field-testing stage, with some of them showing high effectiveness (e.g., *Pseudomonas* spp., *Trichoderma* spp., *Bacillus* spp., non-pathogenic *Fusarium* strains, and arbuscular mycorrhizal fungi) (Bubici et al. 2019). The use of BCAs for controlling FW of bananas is out of the scope of this chapter that rather focuses on the effect of indigenous soil biota development of FW of bananas; however, for those interested in this subject, Bubici et al. (2019) have recently provided a comprehensive detailed and updated revision on this topic. Overall, under field conditions, the FW of bananas has been controlled by up to 79% by using *Pseudomonas* spp. strains and up to 70% by several endophytes and *Trichoderma* spp. strains. Lower biocontrol efficacy (42–55%) has been obtained with arbuscular mycorrhizal fungi, *Bacillus* spp., and non-pathogenic *Fusarium* strains (Bubici et al. 2019).

Other soil biotas, apart from microorganisms, may have a strong influence on the suppressiveness of FW of bananas in certain soils. Thus, modifications in the microbiological properties of the soil caused by the activity of meso- and macro-organisms, such as nematodes, beneficial arthropods, and earthworms, have been identified as relevant factors in the incidence of FW in monoculture banana systems (Shen et al. 2015a). FW of bananas can be exacerbated by the presence of certain nematode species (e.g., *Radopholus similis*) due to root lesions that weaken the plant and facilitate the penetration of the pathogen through the injuries caused by nematode feeding (Dinesh et al. 2014). In the banana agroecosystems, nematodes are one of the main causes of production losses. *R. similis*, *Helicotylenchus* sp., *Pratylenchus coffeae*, *Meloidogyne* sp., and *Rotylenchulus reniformis* are some of the plant-parasitic nematode species frequently associated with crop losses in banana crops (Ritzinger and Fancelli 2006; Almeida et al. 2018).

According to Duyck et al. (2009) and Zhong et al. (2011), banana soils with a high inoculum of *Foc* or FW incidence show a reduced diversity of total nematodes. Thus, under such conditions, the populations of bacterivores (mainly in Rhabditidae,

Pangrolaimidae, and Cephalobidae families), some plant parasites (mainly within Meloidogynidae, Hoplolaimidae, and Pratylenchidae families), and omnivores or predators (mainly in the Qudsianematidae family) decreased in contrast to other groups of nematodes present in non-infested-healthy soils.

3.4.6 Hosts and Dispersion

Plant genetic resistance is generally considered the most plausible strategy and economically feasible measure to effectively manage the FW of bananas (Ploetz 2015a, 2015b). However, resistant cultivars might not match the demands of the market, and the available resistance may be overcome by new pathogenic strains, as was the case of Cavendish and *Foc* TR4 (Dita et al. 2017). Once FW is established in an area, the use of resistant varieties is the most effective, if not the only means to manage the disease. In bananas, complete resistance has only been described in the Cavendish (AAA)-*Foc* R1 interaction. Other interactions, such as Prata (AAB)-*Foc* R1 and Giant Cavendish Tissue Culture Variants (GCTCV)-*Foc* TR4, show intermediate resistance; that is, those banana genotypes develop less severe symptoms than susceptible varieties when grown under similar environmental conditions and *Foc* inoculum levels. However, under-management practices and inoculum pressure are conducive to disease development; FW and yield losses will increase gradually (Dita et al. 2018). Currently, efforts are being made to unravel the genetic and molecular mechanisms driving resistance responses of different banana genotypes using state-of-the-art molecular approaches such as deep-RNA sequencing (Li et al. 2020). A description of the levels of resistance and their implications on FW of banana management is described in Dita et al. (2018).

Foc is a facultative saprophytic fungus with the ability to survive in weeds and grasses. Thus, *Foc* can invade roots of weeds present in banana plantations from hyphae growing from the tissues of senescent banana roots that remain in the soil for long periods (Hennessy et al. 2005). This could explain the persistence of the fungus in soils without banana crops (Ploetz 1994). Thus, the pathogen can infect the roots of certain weeds without causing visible symptoms and can remain in these plants in the absence of a banana crop (Pérez-Vicente et al. 2014). The study of Hennessy et al. (2005) found that the monocotyledonous species *Chloris inflata* and three dicotyledonous species (*Euphorbia heterophylla*, *Tridax procumbens*, and *Cyanthillium cinereum*) can sustain *Foc* inoculum in their root system (Fig. 3.7). Among the different plant species that can be hosts of *Foc*, it is of particular relevance to avoid or control the export of the ornamentals *Canna indica*, *Aglaonema pictum*, and *Hedychium coronarium*, especially in areas close to FW-affected fields, since these species can serve as alternative hosts for the pathogen (Waman et al. 2013). Additionally, it is known that monocotyledons are excellent endophytic carriers of *F. oxysporum* (Deltour et al. 2017) and could contribute to the increase or maintenance of *Foc* inoculum.

Fig. 3.7 Main host plants (type, family, and species) of *Fusarium oxysporum* f. sp. *cubense*

The systemic growth of *Foc* in xylem tissues of infected asymptomatic banana plants represents one of the primary ways in which the fungus can be introduced into a free pathogen growing area (Ploetz 2015a). The influence of the anthropogenic component by the active movement of infected planting material, equipment, and people between infected and not affected areas had important repercussions in the dissemination of *Foc* R1 and TR4 (Dita et al. 2018). According to Stover (1990), the epidemic of *Foc* R1 in "Gros Michel" was due to the absence of quarantine measures, the use of infected planting material, and the establishment of new plantations using machinery and propagation material from infected fields (Dita et al. 2018). Thus, the pathogen is dispersed by the movement of propagating material and infested agricultural tools that have contaminated soil, runoff water during rainy events, floods, and irrigation (Ploetz 2015a). With the rain, the spores of the pathogen, as well as the infected material, are transported to the drainage channels, and through irrigation water, these spores infect new areas.

3.5 Risk Analysis of *Fusarium oxysporum* f. sp. *cubense* Occurrence in Banana Plantations in Venezuela

Venezuela has a diverse ecology, and it is divided into several agroecological zones (AEZs). This division in AEZs is based on the geographic location, edaphic and climatic characteristics, agricultural potential, and predominant agricultural production systems (INIA 2016). Venezuela has a hot and humid tropical climate that is more moderate in the highlands. Caracas, the capital, has a tropical savanna climate. The average annual temperature ranges between 24 °C and 27 °C (Fig. 3.8a). Annual precipitation varies between 400 mm in arid zones and 4000 mm in the Venezuelan Amazon (Fig. 3.8b).

The delimitation of the main cultivation areas of Musaceae in Venezuela and the total yield of the area using the Spatial Production Allocation Model (SPAM) was established (Wood-Sichra et al. 2016), developed by the International Food Policy Research Institute (IFPRI), at a spatial resolution of 5 arc-minutes (approximately 10 km^2) (Fig. 3.9a) (You et al. 2014).

The designated areas with a different potential risk of *Foc* occurrence in the main banana-producing areas in Venezuela were based on the assembly of agro-environmental factors (climate, soil type, and agronomic management) (Fig. 3.9a). Additionally, these analyses were complemented by five Musaceae experts of Red Venezolana de Musáceas (Musaven), who classified production systems with different levels of risk or susceptibility (high-, moderate-, and low-risk areas), according to the characteristics established by the FAO (Pérez-Vicente et al. 2014).

The classification was based on the prevailing characteristics of the different Musaceae production systems in Venezuela with four important steps, including risk identification (sources, communities, and production systems), analysis (probability and consequences), evaluation (prioritization), and assessment (response to the risk). Our results defined three areas with different potential risks of *Foc* occurrence in Venezuela (high-risk areas, moderate to high-risk areas, and moderate to low-risk areas).

3.5.1 *High-Risk Areas*

Cavendish banana large-scale farms located in Colón, Francisco Javier Pulgar, and La Cañada de Urdaneta in the Zulia state are at high potential risk (Fig. 3.9b) because (i) they include large plantations (500–1900 ha) of genetic clones highly susceptible to *Foc* TR4 like Cavendish (Pineo Gigante, Williams); (ii) they present edaphoclimatic characteristics (precipitation greater than 1200 mm and a mean diurnal temperature range of 24–34 °C) suitable for the appearance and

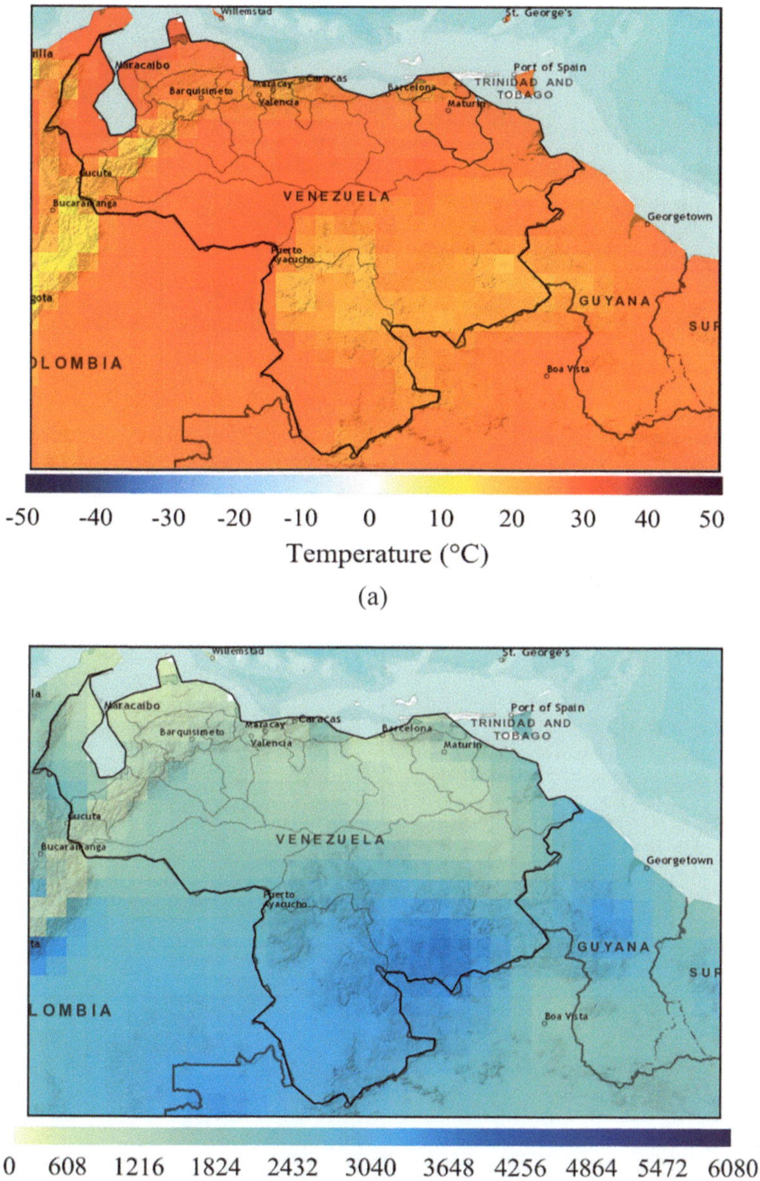

(b)

Fig. 3.8 (**a**) Observed climatology of (**a**) mean-temperature (1991–2020); (**b**) precipitation (1991–2020) in Venezuela according to the data source from Harris et al. (2020)

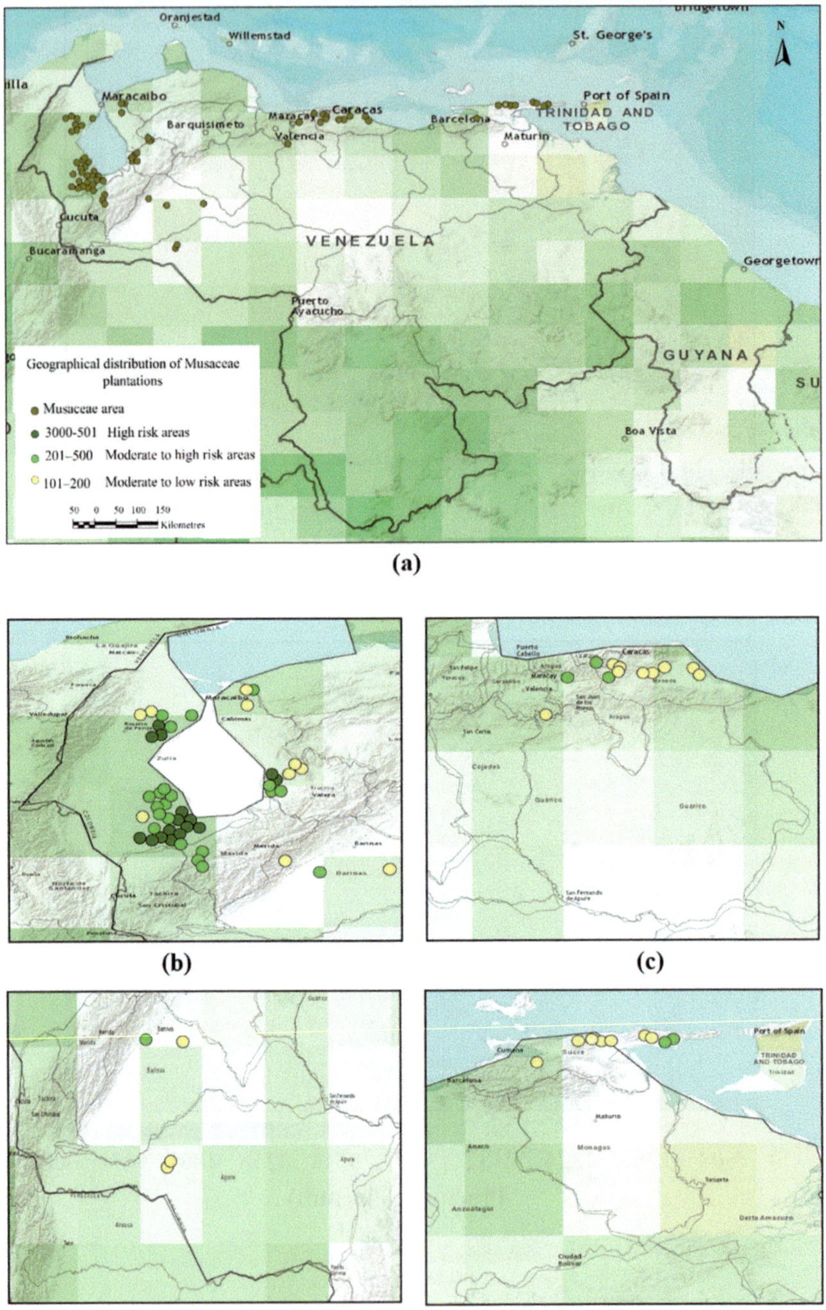

Fig. 3.9 (**a**) Total area of Musaceae production in Venezuela according to the USDA (2022) and potential risk of *Fusarium oxysporum* f. sp. *cubense* occurrence in Venezuela in (**b**) the western region, (**c**) central region, (**d**) southwestern region, and (**e**) eastern region

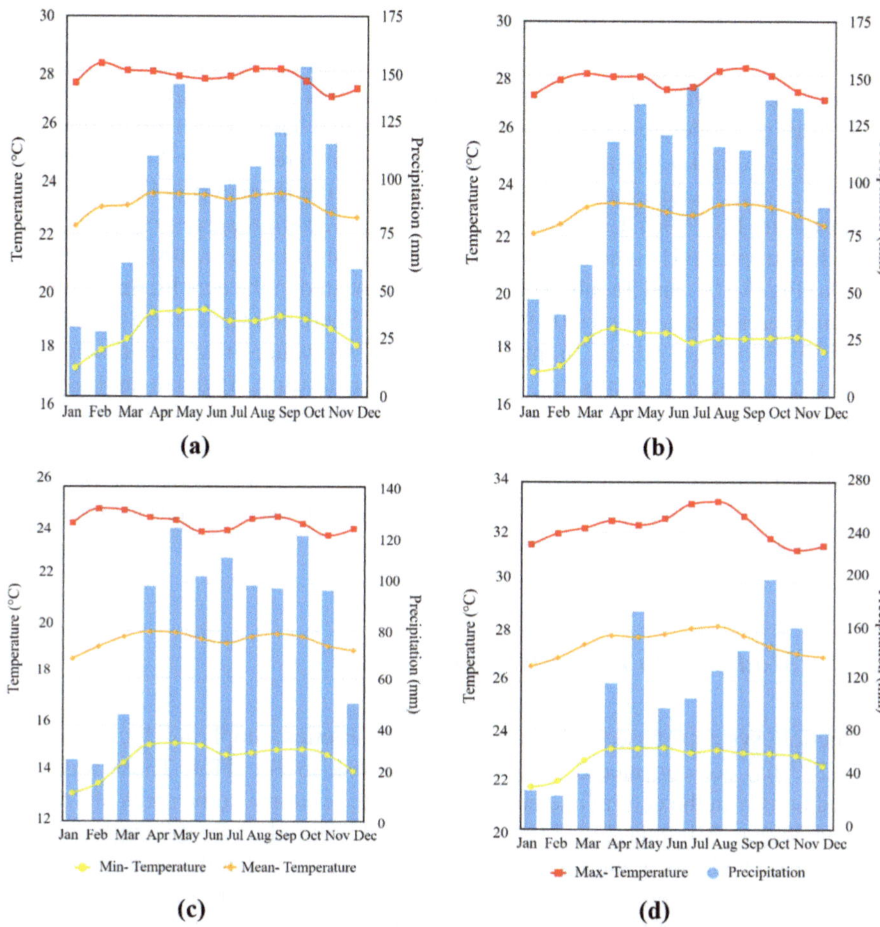

Fig. 3.10 Monthly climatology of mean temperature, minimum temperature, maximum temperature, and precipitation (1991–2020) in the (**a**) Trujillo, (**b**) Tachira, (**c**) Mérida, and (**d**) Zulia states according to the data source from Harris et al. (2020)

establishment of the pathogen (Fig. 3.10d); (iii) they implement intensive production systems, which requires the continuous movement of employees between plantations; and (iv) a single source of water for irrigation can favor the spread of the fungus within and across plantations on soil particles and irrigation water. Additionally, the physical proximity to Colombia puts these western municipalities of Venezuela at risk because the border of Venezuela is approximately 120 km from the *Foc*-infected area in Colombia and at a distance of around 300 km from the banana zones to the south of Lake Maracaibo.

3.5.2 Moderate to High-Risk Areas

Medium-scale Cavendish banana production systems (200–500 ha) can be considered areas of moderate to high vulnerability to *Foc* TR4. These plantations are mainly located in La Guajira in the Zulia state, some areas in the Alberto Adriani and Sucre municipalities in the State of Mérida and the Barinas municipality in the Barinas state (Fig. 3.9d). La Guajira is located in the ecoclimatic region of dry areas of the low tropics (G1) whose average annual rainfall is between 400 and 600 mm. Mérida (Fig. 3.10c) and Tachira (Fig. 3.10b) are located in the ecoclimatic region of the humid premontane tropics (D1) as well as the Sucre and Candelaria municipalities in Trujillo (Fig. 3.10a). Its main limitations are water erosion and acidity of soils, which coexist with some areas of better fertility. Pedraza and Barinas Municipalities in the Barinas state are located in the ecoclimatic region of humid tropics low (B2); they are areas with varied topography and soils with low to very low natural fertility. As indicated for large-scale production systems, the pathogen can enter and disseminate due to the movement of employees, materials, tools, and irrigation water within and between farms.

3.5.3 Moderate to Low-Risk Areas

Production systems based on mixed banana varieties supplying local markets (100–200 ha) are considered moderate to low risk to *Foc* TR4. These areas are located mainly in the central and eastern parts of the country, some areas of Trujillo, Aragua, Vargas, Miranda (Fig. 3.9c), Sucre (Fig. 3.9e), and south of Lake Maracaibo (Fig. 3.9b). The Bolivar and La Ceiba municipalities in the Trujillo state are located in the ecoclimatic region of the subhumid tropics low (A1). The Brion and Buroz municipalities in the Miranda state (Fig. 3.11b) and the Libertador municipality in the Aragua state are located in the ecoclimatic region of the subhumid tropics low (A1) (Fig. 3.11c). The La Guajira state is located in the ecoclimatic region of the subhumid tropics low (A4) (Fig. 3.11d); they are flat areas, with low to very low natural soil fertility and excessive tendency drainage. The Andrés Eloy Blanco and Andrés Mata municipalities in the Sucre state are located in the ecoclimatic region of the subhumid premontane tropics (C1) (Fig. 3.11a). Paéz Municipality in the Apure state (Fig. 3.12a) and Obispos Municipality in the Barinas state are located in the ecoclimatic region of the subhumid tropics low (A5) (Fig. 3.12b). They are flat areas with good to moderate drainage soils and moderate fertility. This risk category also includes production systems based on intercropping.

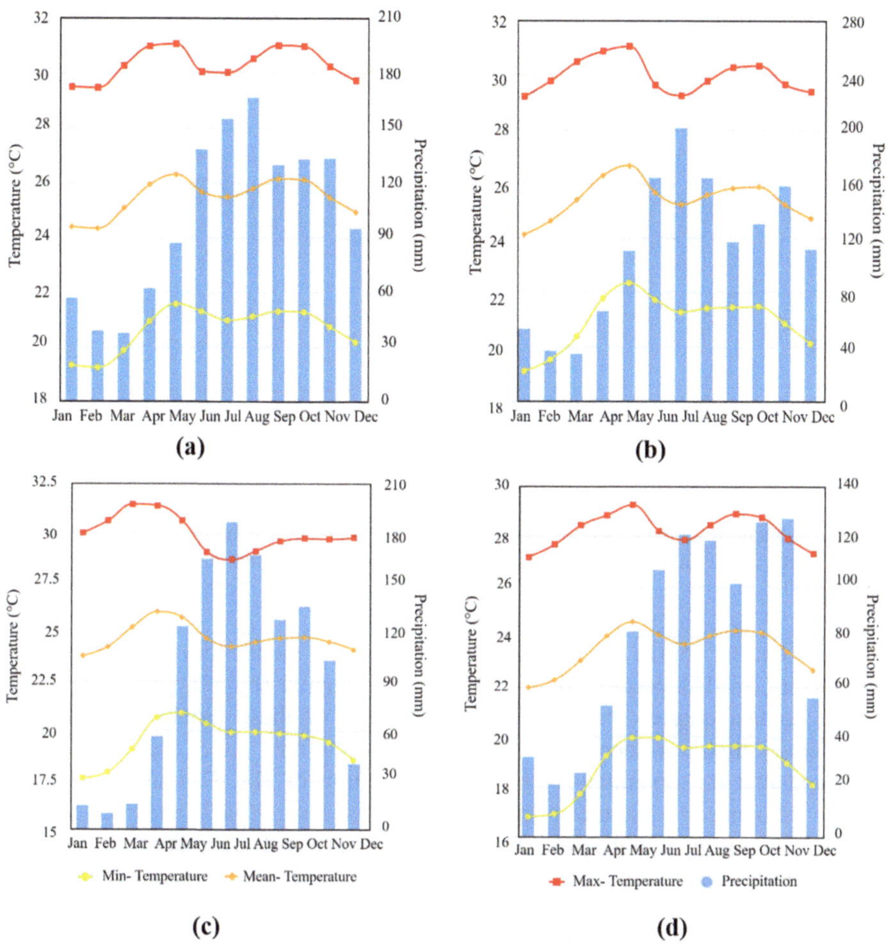

Fig. 3.11 Monthly climatology of temperature mean, minimum, maximum, and precipitation (1991–2020) in the (**a**) Sucre, (**b**) Miranda, (**c**) Aragua, and (**d**) La Guajira states according to the data source from Harris et al. (2020)

3.6 Climatic Suitability for *Foc* TR4 Occurrence in Venezuela

Multicollinearity was addressed by the variance inflation factor (VIF) using a threshold of 10 (Hijmans et al. 2005). The climate suitability was estimated based on bioclimatic variables representing annual trends. Nine non-correlated bioclimatic variables were selected based on VIF for model fitting. These variables included the mean diurnal range (mean monthly maximum temperature-minimum temperature) (bio2), isothermality (bio3), mean temperature of the wettest quarter (bio8), mean temperature of the driest quarter (bio9), mean temperature of the

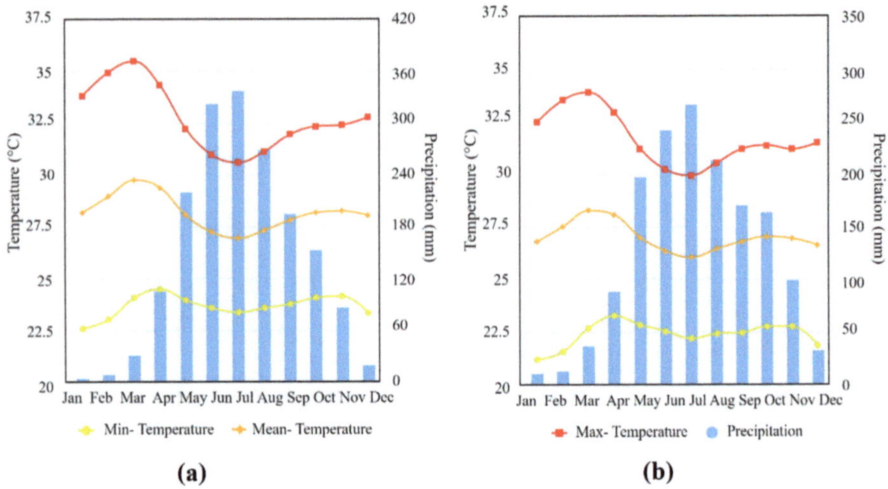

Fig. 3.12 Monthly climatology of temperature mean, minimum, maximum, and precipitation (1991–2020) in the (**a**) Apure and (**b**) Barinas states according to the data source from Harris et al. (2020)

warmest quarter (bio10), precipitation of the coldest month (bio13), precipitation seasonality (bio15), precipitation of the warmest quarter (bio18), and precipitation of the coldest quarter (bio19). The contribution of these bioclimatic variables assessed by their variance importance based on the area under the curve (AUC) indicated that the potential geographic distribution of *Foc* TR4 in Venezuela is mostly influenced by the mean diurnal temperature range (bio2) and precipitation, particularly during extreme cold (bio13) and warm (bio18) periods of the season.

Results indicated that the performance of the model was considered high. The area under the curve (AUC) of the receiver operating characteristic (ROC) was estimated in 0.93 ± 0.0236, the sensitivity in 0.94 ± 0.052, the specificity in 0.87 ± 0.044, and the true skill statistics (TSS) and the Cohen's kappa were estimated in 0.81 ± 0.052 and 0.55 ± 0.085, respectively. The map with the continuous suitability scores for *Foc* TR4 in Venezuela drawn from the MaxEnt model is presented in Fig. 3.13a and b show four-level categorized suitability. Except for some areas in the central and eastern parts of the Bolivar state, the northern part of the Amazonas state, and the Andes Mountains in the west, for which the potential for establishment is estimated negligible, the rest of the country could have climatic conditions that would allow the establishment of *Foc* TR4, with the central part of the country being classified with low suitability between 0.1 and 0.3; interestingly, all regions in northern states are classified as moderately suitable between 0.3 and 0.6, and most of the banana-producing areas in the Zulia, Trujillo, Miranda, and Sure states have climatic conditions that are estimated as highly suitable (>0.6) for the establishment of *Foc* TR4 (Fig. 3.13a). In the current climate conditions, the areas of great commercial importance of the La Cañada de Urdaneta and La Guajira municipalities in the Zulia state show the highest suitability scores (>0.90), followed by the Buroz

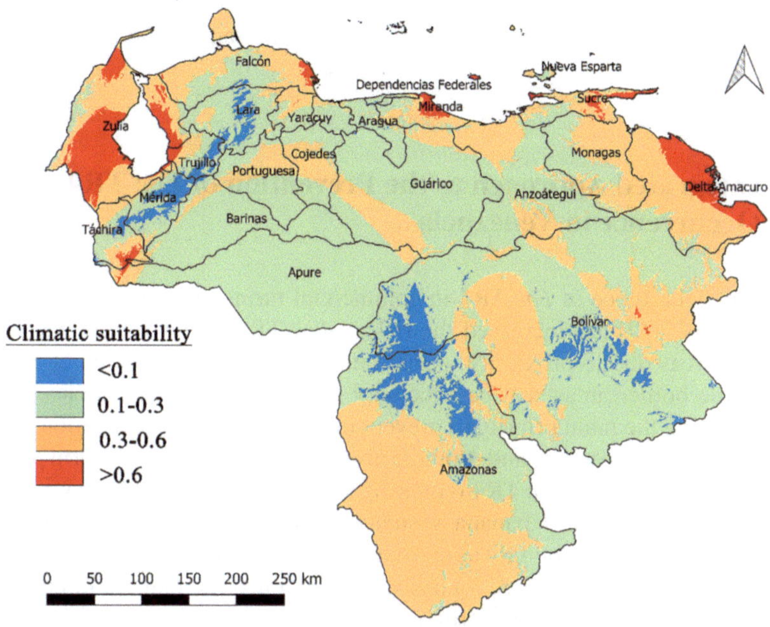

(a)

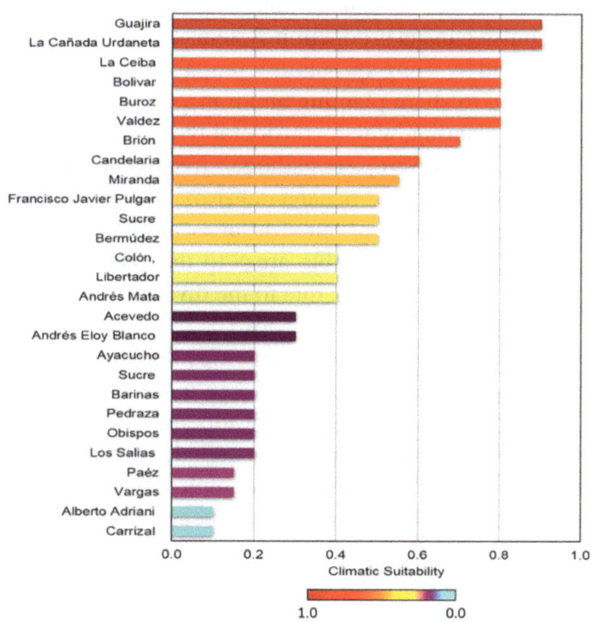

(b)

Fig. 3.13 Estimated climatic suitability maps for *Fusarium oxysporum* f. sp. *cubense* TR4 according to a maximum entropy model. (**a**) Climatic suitability index is categorized into four levels: unsuitable (<0.1), low (0.1–0.3), moderate (0.3–0.6), and highly suitable (>0.6) (Olivares et al. 2021). (**b**) General approximation of the continuous scores of climatic suitability for Foc TR4 in the banana municipalities of Venezuela

(0.80) and Brión (0.70) municipalities in the Miranda state, as well as the banana areas of the Trujillo state, specifically in the Bolivar and La Ceiba municipalities (0.80) (Fig. 3.13b).

3.7 Integrated Approach to the Prevention of *Foc* TR4 Occurrence in Venezuela

The recent occurrence of *Foc* TR4 in commercial farms in the Colombian Guajira region in July 2019 has increased the risk of *Foc* TR4 in the Venezuelan banana-producing areas due to proximity to the outbreak. According to Dita et al. (2013, 2017), these border areas will be necessary to implement state policies, with the exclusion being the main priority to prevent the movement of propagation material, establish quarantine sectors, and promote awareness campaigns to inform producers of the serious threat that *Foc* TR4 represents for the Venezuelan banana sector.

In Venezuela, intensive banana systems located in the western region (Zulia state) are more at risk of *Foc* TR4 as extremely strict biosafety protocols are required. Due to the absence of efficient control measures, once the plant is infected in the field, the best way to protect the crop is to prevent the introduction of the pathogen from affected areas and the use of resistant banana cultivars to *Foc* TR4 such as Giant Cavendish tissue-culture variants (Huang et al. 2005).

Small-scale Cavendish banana production systems located in most of Venezuela's territory are characterized by using sprouts as a propagation system instead of tissue crops. Farmers also tend to share planting and transport tools and neglect sanitary practices at the time of planting, which considerably increases the chance of introducing the fungus from affected crops to pathogen-free plantations. On the contrary, production systems based on mixed banana cultivars, as well as mixed crops, subsistence, and indigenous farming systems in areas of the southwest and east of Venezuela are considered less vulnerable areas to the economic damage that the occurrence of *Foc* TR4 pathogen attack due to the diversity of banana cultivars and the possibility to replace banana by other crops.

In this context, Fig. 3.14 lists the main activities that could be implemented to prevent the entry and spread of this pathogen and more specifically of the highly virulent *Foc* TR4 in Venezuela. In summary, results indicate that the best approach to fight against the threat that *Foc* TR4 assumes includes prevention approaches, the use of resistant/tolerant cultivars, as well as the implementation of good phytosanitary practices at planting time. However, these measures are very expensive and require highly trained personnel to identify the pathogen and control it. Unfortunately, in areas still free of this highly virulent race, such as the case of Venezuela, there are few specialists working on this disease. Consequently, it is important to build technical capabilities in growers, service providers, and national plant protection agents to identify *Foc* symptoms in the field and to implement identification methodologies

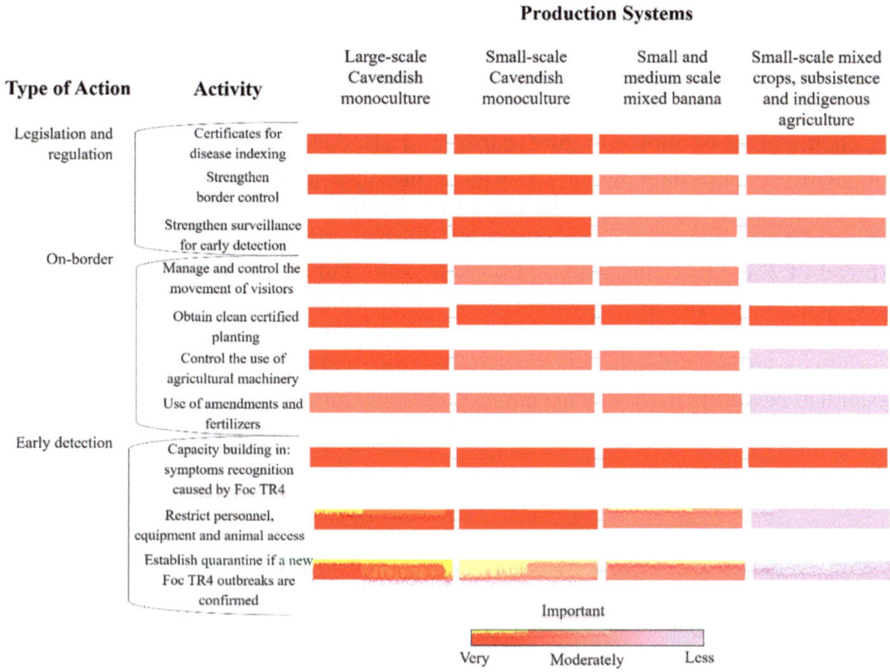

Fig. 3.14 Activities aimed at preventing the entry and spread of *Fusarium oxysporum* f. sp. *cubense* (*Foc*) TR4 in the banana production systems of Venezuela. (Adapted from the FAO (Pérez-Vicente et al. 2014; Martinez et al. 2023))

based mainly on the use of modern molecular diagnostic tools to unequivocally identify this highly virulent race of the pathogen.

3.8 Conclusions

There are several lines to follow after this new situation of the detection of *Foc* TR4 in Latin America and the Caribbean since 2019, since the threat that *Foc* TR4 poses to agriculture must be addressed from different angles, without losing sight of any focus. The *Foc* TR4 prevention and containment programs that are currently available must be complied with, but, at the same time, it is necessary, as previously mentioned, to advance in the knowledge of the multiple factors related to this situation, since in this way, it will be possible to design management strategies for this disease based on scientific knowledge.

Despite the scientific progress in recent years of *Foc* in tropical regions, there are still not enough studies to define the prevention and control mechanisms of this disease that is so dangerous for bananas. The correlations between the incidence and virulence of *Foc* with different climatic variables, soil properties, and management practices are not clear, despite the literature reported in this chapter, although it allows a more comprehensive understanding of the impact of these properties and their interactions. Today, the scientific and technological progress in terms of pathogens such as *Foc* has not been enough to define a product (biological or chemical, or of another nature) capable of effectively eradicating the disease in the field, although there are efforts with certain biological control agents.

In the case of Venezuela, the use of eradication or exclusion protocols in the affected areas is proposed as one of the most important measures to prevent the spread of *Foc* TR4. The factors of climate, soil, and agronomic management presented in this chapter provide an overview of their influence on the susceptibility of banana plants to *Foc*. This information collected in this chapter could improve agricultural practices in banana plantations, promote soil management alternatives, and establish new farm locations in suitable areas, which will generate more opportunities to control the coexistence between *Foc* and banana production in Venezuela.

References

Aguilar EA, Turner DW, Sivasithamparam K (2000) Fusarium oxysporum f. sp cubense inoculation and hypoxia alter peroxidase and phenylalanine ammonia lyase activities in nodal roots of banana cultivars (Musa sp.) differing in their susceptibility to Fusarium wilt. Aust J Bot 48:589–596. https://doi.org/10.1071/bt99009

Almeida NO, Teixeira RA, Carneiro FA, de Oliveira CM, Ribeiro VA, Lobo Junior M, da Rocha MR (2018) Occurrence and correlations of nematodes, fusarium oxysporum and edaphic factors on banana plantations. J Phytopathol 166:265–272. https://doi.org/10.1111/jph.12683

Bentley S, Pegg KG, Moore NY, Davis RD, Buddenhagen IW (1998) Genetic variation among vegetative compatibility groups of *Fusarium oxysporum* f. sp. *cubense* analyzed by DNA fingerprinting. Phytopathology 88:1283–1293. https://doi.org/10.1094/phyto.1998.88.12.1283

Borges A, Angulo Rodríguez D, Gutiérrez Jerez F, Trujillo I (1983) Estudio sobre el Mal de Panamá en las Islas Canarias. II. Influencia de los desequilibrios nutritivos P-Zn y K-Mg del suelo, en la alteración de los mecanismos de resistencia de la platanera (Cavendish enana) al Mal de Panamá. Fruits 38:755–758

Borges AF, Bravo JJ, Perez JF, Lopez I (1991) Enhanced resistance of banana plants (dwarf Cavendish) to *fusarium oxysporum* f. sp. *cubense* by controlled Zn nutrition under field conditions. Banan Newsl 14:24–26

Bosman M (2016) Role of the environment on the incidence of Panama disease in bananas. Wageningen University, Wageningen, The Netherlands, pp 12–68

Brake VM, Pegg KG, Irwin JAG, Chaseling J (1995) The influence of temperature, inoculum level and trace of fusarium-oxysporum f sp cubense on the disease reaction of banana cv cavendish. Aust J Agric Res 46:673–685. https://doi.org/10.1071/ar9950673

Bronick CJ, Lal R (2005) Soil structure and management: a review. Geoderma 124:3–22. https://doi.org/10.1016/j.geoderma.2004.03.005

Bubici G, Kaushal M, Prigigallo MI, Cabanas CGL, Mercado-Blanco J (2019) Biological control agents against fusarium wilt of banana. Front Microbiol 10. https://doi.org/10.3389/fmicb.2019.01290

Buddenhagen I (2009) Understanding strain diversity in fusarium oxysporum f. sp cubense and history of introduction of 'tropical race 4' to better manage Banana production. Acta Hortic 828:193–204. https://doi.org/10.17660/ActaHortic.2009.828.19

CABI/EPPO (2015) Fusarium oxysporum f. sp. cubense. Distribution Maps of Plant Diseases, Map 31, 7th edn. CABI, Wallingford, UK

Chuang TY (1991) Soil suppressive of banana fusarium wilt in Taiwan. Plant Prot Bull 33:133–141

Cortez A, Olivares B, Parra R, Lobo D, Rodríguez MF, Rey JC (2018) Descripción de los eventos de sequía meteorológica en localidades de la cordillera central, Venezuela. Cs Ing Apls I(1):22–44. https://doi.org/10.22206/cyap.2018.vlil.pp23-45

Deltour P, Franca SC, Pereira OL, Cardoso I, De Neve S, Debode J, Hofte M (2017) Disease suppressiveness to fusarium wilt of banana in an agroforestry system: influence of soil characteristics and plant community. Agric Ecosyst Environ 239:173–181. https://doi.org/10.1016/j.agee.2017.01.018

Deng X, Li QF, Hou XW, Wu CY (2013) Soil microbial functional diversity from different infection grades of banana fusarium wilt (*fusarium oxysporum* f. sp *cubense*). Appl Mech Mater 295:2274–2280. https://doi.org/10.4028/www.scientific.net/AMM.295-298.2274

Dinesh BM, Ravichandra NG, Reddy BMR, Somasekhara YM (2014) Interactions between *Radopholus similis* and *fusarium oxysporum* f. sp. *cubense* causing wilt Complex on Banana. Int J Adv Res 2:976–985

Dita M, Echegoyén Ramos PE, Pérez Vicente LF (2013) Plan de Contingencia Ante un Brote de la Raza 4 Tropical de *Fusarium oxysporum* f. sp. *cubense*, en un País de la Región Del OIRSA. OIRSA Editorial, San Salvador, El Salvador, p 155

Dita M, Perez-Vicente L, Guzman M, Urias C, Staver C (2017) Fusarium tropical race 4 and it threat to food security and the banana industry in Latin America and the Caribbean. Phytopathology 107:8–8

Dita M, Barquero M, Heck D, Mizubuti ESG, Staver CP (2018) Fusarium wilt of banana: current knowledge on epidemiology and research needs toward sustainable disease management. Front Plant Sci 9. https://doi.org/10.3389/fpls.2018.01468

Domínguez J, Negrin MA, Rodriguez CM (2001) Aggregate water-stability, particle-size and soil solution properties in conducive and suppressive soils to fusarium wilt of banana from Canary Islands (Spain). Soil Biol Biochem 33:449–455. https://doi.org/10.1016/s0038-0717(00)00184-x

Domínguez J, Negrin MA, Rodriguez CM (2008) Soil potassium indices and clay-sized particles affecting banana-wilt expression caused by soil fungus in banana plantation development on transported volcanic soils. Comm Soil Sci Plant Anal 39:397–412. https://doi.org/10.1080/00103620701826522

Duyck PF, Pavoine S, Tixier P, Chabrier C, Queneherve P (2009) Host range as an axis of niche partitioning in the plant-feeding nematode community of banana agroecosystems. Soil Biol Biochem 41:1139–1145. https://doi.org/10.1016/j.soilbio.2009.02.020

EPPO (2014) PQR database. Paris, France: European and Mediterranean Plant Protection Organization https://gdeppoint/taxon/FUSACB/distribution Accessed 20 Feb 2020

FAO (2019) Banana Statistical Compendium year 2019. http://www.fao.org/3/cb0466en/cb0466en.pdf. Accessed 16 March 2020

FONTAGRO (2021) Fusarium R4. https://wwwfontagroorg/new/proyectos/fusarium-R4 Accessed 16 March 2021

Garcia-Bastidas F, Ordonez N, Konkol J, Al-Qasim M, Naser Z, Abdelwali M, Salem N, Waalwijk C, Ploetz RC, Kema GHJ (2014) First report of fusarium oxysporum f. sp cubense tropical race 4 associated with Panama disease of banana outside Southeast Asia. Plant Dis 98:694–694. https://doi.org/10.1094/pdis-09-13-0954-pdn

Ghag SB, Shekhawat UKS, Ganapathi TR (2015) Fusarium wilt of banana: biology, epidemiology and management. Int J Pest Manag 61:250–263. https://doi.org/10.1080/09670874.2015.1043972

Groenewald S, van den Berg N, Marasas WFO, Viljoen A (2006) Biological, physiological and pathogenic variation in a genetically homogenous population of *fusarium oxysporum* f. sp. *cubense*. Australas Plant Pathol 35:401–409. https://doi.org/10.1071/AP06041

Gutiérrez F, Borges-Pérez A, Trujillo I (1983) Estudio sobre el mal de Panamá en las islas canarias. I Características físicas y químicas de los suelos y su relación con la aparición de la enfermedad Fruits 38:677–682

Harris I, Osborn TJ et al (2020) Version 4 of the CRU TS monthly high-resolution gridded multivariate climate dataset. Sci Data 7:109. https://doi.org/10.1038/s41597-020-0453-3

Hennessy C, Walduck G, Daly A, Padovan A (2005) Weed hosts of *fusarium oxysporum* f. sp *cubense* tropical race 4 in northern Australia. Australas Plant Pathol 34:115–117. https://doi.org/10.1071/ap04091

Hijmans RJ, Cameron SE, Parra JL, Jones PG, Jarvis A (2005) Very high-resolution interpolated climate surfaces for global land areas. Int J Climatol 25:1965–1978. https://doi.org/10.1002/joc.1276

Huang B, Xu L, Molina AB Jr (2005) Preliminary evaluation of IMTP-III varieties and local cultivars against fusarium wilt disease in South China. Rome, Italy, International Plant Genetic Resources Institute (IPGRI), pp 187–192

Huang X, Wen T, Zhang J, Meng L, Zhu T, Cai Z (2015) Toxic organic acids produced in biological soil disinfestation mainly caused the suppression of fusarium oxysporum f. sp cubense. BioControl 60:113–124. https://doi.org/10.1007/s10526-014-9623-6

Huber D, Römheld V, Weinmann M (2012) Relationship between nutrition, plant diseases and pests. In: Marschner P (ed) Mineral. Nutrition of higher plants, 3rd edn. Academy Press, Stuttgart, Germany, pp 283–298. https://doi.org/10.1016/B978-0-12-384905-2.00010-8

INIA (Instituto Nacional de Investigaciones Agrícolas, VE) (2016) Aproximación agroecológica para el nuevo modelo de producción agrícola en Venezuela. Editorial del Instituto Nacional de Investigaciones Agrícolas, Maracay, Venezuela, pp 12–90

Kema GHJ, Drenth A, Dita M, Jansen K, Vellema S, Stoorvogel JJ (2021) Editorial: fusarium wilt of banana, a recurring threat to global banana production. Front Plant Sci 11:1–6

Koberl M, Dita M, Martinuz A, Staver C, Berg G (2015) Agroforestry leads to shifts within the gammaproteobacterial microbiome of banana plants cultivated in Central America Front Microbiol 6. https://doi.org/10.3389/fmicb.2015.00091

Koberl M, Dita M, Martinuz A, Staver C, Berg G (2017) Members of Gammaproteobacteria as indicator species of healthy banana plants on fusarium wilt-infested fields in Central America. Sci Rep 7(9). https://doi.org/10.1038/srep45318

Lahav E, Israeli Y (2000) Mineral deficiencies of Banana. In: Jones D (ed) Diseases of banana, Abacá and enset. CABI Publishing, Wallingford, pp 339–350

Li Z, Deng Z, Chen S, Yang H, Zheng Y, Dai L, Zhang F, Wang S, Hu S (2018) Contrasting physical and biochemical properties of orchard soils suppressive and conducive to fusarium wilt of banana. Soil Use Manag 34:154–162. https://doi.org/10.1111/sum.12390

Li WM, Dita M, Rouard M, Wu W, Roux N, Xie JH, Ge XJ (2020) Deep RNA-seq analysis reveals key responding aspects of wild banana relative resistance to *fusarium oxysporum* f. sp. *cubense* tropical race 4. Funct Integr Genom 20:551–562. https://doi.org/10.1007/s10142-020-00734-z

Martelo MT (2003) Metodología para la selección de Modelos de Circulación General de la Atmósfera y Escenarios Climáticos a incluir en la Primera Comunicación Nacional en Cambio Climático de Venezuela, MARN. Caracas, Venezuela, 51 p

Martínez GE, Rey-Brina JC, Rodríguez D, Jiménez C, Rodríguez Y, Rumbos R, Pargas-Pichardo R, Martínez E (2020) Análisis de la situación fitopatológica actual de las musáceas comestibles en Venezuela. Agr Trop 70:1–20. https://doi.org/10.5281/4323273

Martínez G, Olivares BO, Rey JC, Rojas J, Cardenas J, Muentes C, Dawson C (2023) The Advance of Fusarium Wilt Tropical Race 4 in Musaceae of Latin America and the Caribbean: Current Situation. Pathogens 12(2):277. https://doi.org/10.3390/pathogens12020277

Maryani N, Lombard L, Poerba YS, Subandiyah S, Crous PW, Kema GHJ (2009) Phylogeny and genetic diversity of the banana fusarium wilt pathogen fusarium oxysporum f. sp. cubense in the Indonesian Centre of origin. Stud Mycol 92:155–194. https://doi.org/10.1016/j.simyco.2018.06.003

Maryani N, Lombard L, Poerba YS, Subandiyah S, Crous PW, Kema GHJ (2018) Phylogeny and genetic diversity of the banana fusarium wilt pathogen *fusarium oxysporum* f. sp. *cubense* in the Indonesian Centre of origin. Stud Mycol 91(1):79–99

Ministerio del Ambiente y de los Recursos Naturales (MARN) (2005) Primera Comunicación Nacional en Cambio Climático de Venezuela. Programa de las Naciones Unidas para el Desarrollo (PNUD), Caracas, p 134

Molina AB, Fabregar E, Sinohin VG, Yi G, Viljoen A (2009) Recent occurrence of fusarium oxysporum f. sp cubense tropical race 4 in Asia. Acta Hortic 828:109–115. https://doi.org/10.17660/ActaHortic.2009.828.10

Nansamba M, Sibiya J, Tumuhimbise R, Karamura D, Kubiriba J, Karamura E (2020) Breeding banana (*Musa* spp.) for drought tolerance: a review. Plant Breed 139:685–696

Nasira N, Pittaway PA, Pegg KG (2003) Effect of organic amendments and solarization on fusarium wilt in susceptible banana plantlets, transplanted into naturally infested soil. Aust J Agric Res 54:251–257. https://doi.org/10.1071/ar02099

Nelson PE (1991) History of fusarium systematics. Phytopathology 81:1045–1048

Olivares BO (2018) Tropical rainfall conditions in rainfed agriculture in Carabobo, Venezuela. Lgr 27:86–102. https://doi.org/10.17163/lgr.n27.2018.07

Olivares B, Zingaretti ML (2018) Analysis of the meteorological drought in four agricultural localities of Venezuela through the combination of multivariate methods. UNED Research Journal 10(1):181–192

Olivares B, Zingaretti ML (2019) Aplicación de métodos multivariados para la caracterización de periodos de sequía meteorológica en Venezuela. Revista Luna Azul 48(172):192. https://doi.org/10.17151/luaz.2019.48.10

Olivares B, Cortez A, Lobo D, Parra R, Rey J, Rodriguez M (2017a) Evaluation of agricultural vulnerability to drought weather in different locations of Venezuela. Rev Fac Agron (LUZ) 34:103–129

Olivares B, Parra R, Cortez A (2017b) Characterization of precipitation patterns in Anzoátegui state, Venezuela. Ería 3(3):353–365. https://doi.org/10.17811/er.3.2017.353-365

Olivares BO, Hernández R, Coelho R, Molina JC, Pereira Y (2018) Analysis of climate types: Main strategies for sustainable decisions in agricultural areas of Carabobo, Venezuela. Sci agropecu 9:359–369. https://doi.org/10.17268/sci.agropecu.2018.03.07

Olivares BO, Araya-Alman M, Acevedo-Opazo C, Rey J.C, Cañete-Salinas P, Kurina FG, Balzarini M, Lobo D, Navas-Cortés JA, Landa BB (2020) Relationship between soil properties and banana productivity in the two main cultivation areas in Venezuela. J Soil Sci Plant 20: 2512–2524. https://doi.org/10.1007/s42729-020-00317-8

Olivares BO, Rey JC, Lobo D, Navas-Cortés JA, Gómez JA, Landa BB (2021) Fusarium wilt of bananas: a review of agro-environmental factors in the Venezuelan production system affecting its development. Agronomy 11(5):986. https://doi.org/10.3390/agronomy11050986

Ovalles F, Cabrera-Bisbal E, Cortez A, Rodríguez M, Rey JC, Comerma J (2005) Aproximación a los escenarios de adaptación al cambio climático del sector agrícola. Maracay, INIA, p 146

Ovalles F, Cortez A, Rodríguez M, Rey JC, Cabrera-Bisbal E (2008) Variación geográfica del impacto del cambio climático en el sector agrícola en Venezuela. Agronomía Trop 58(1):37–40

Pattison AB, Wright CL, Kukulies TL, Molina AB (2014) Ground cover management alters development of fusarium wilt symptoms in Ducasse bananas. Australas Plant Pathol 43:465–476. https://doi.org/10.1007/s13313-014-0296-5

Pegg KG, Moore NY, Bentley S (1996) Fusarium wilt of banana in Australia: a review. Aust J Agric Res 47:637–650. https://doi.org/10.1071/ar9960637

Peng HX, Sivasithamparam K, Turner DW (1999) Chlamydospore germination and fusarium wilt of banana plantlets in suppressive and conducive soils are affected by physical and chemical factors. Soil Biol Biochem 31:1363–1374. https://doi.org/10.1016/S0038-0717(99)00045-0

Pérez L, Batlle A, Fonseca J (2003) Fusarium oxysporum f. sp. cubense en Cuba: Biología de las poblaciones, reacción de los clones híbridos de la FHIA y biocontrol. In Proceedings of the Taller Internacional manejo convencional y alternativo de la Sigatoka negra, nematodos y otras plagas asociadas al cultivo de Musáceas, MUSALAC/INIBAP, Guayaquil, Ecuador, 11–13 August 2003

Pérez-Vicente L (2004) Marchitamiento por Fusarium (Mal de Panamá) en bananos: Una revisión actualizada del conocimiento presente sobre su agente causal. Fitosanidad 8:27–38

Pérez-Vicente L, Porras Á (2015) Impacto potencial del cambio climático sobre las plagas de bananos y plátanos en Cuba. Fitosanidad 19:201–211

Pérez-Vicente L, Dita MA, de la Parte Martínez E (2014) Technical Manual: Prevention and diagnostic of Fusarium Wilt (Panama disease) of banana caused by *Fusarium oxysporum* f. sp. cubense Tropical Race 4 (TR4). Prepared for the FAO/CARDI Regional Workshop on the Diagnosis of Fusarium Wilt (Panama disease) caused by *Fusarium oxysporum* f. sp. *cubense* Tropical Race 4: Mitigating the Threat and Preventing it Spread in the Caribbean. University of West Indies, Trinidad and Tobago, 2014

Pirttila AM (2016) Commentary: Agroforestry leads to shifts within the gammaproteobacterial microbiome of banana plants cultivated in Central. America Front Microbiol:7. https://doi.org/10.3389/fmicb2016.006.56

Ploetz RC (1990) Population biology of Fusarium oxysporum f sp cubense. In Fusarium Wilt of Banana; Ploetz, RC, Ed; APS Press: St Paul, Minnesota, USA 1:63–76

Ploetz RC (1994) Panama-disease—return of the first banana menace. Int J Pest Manag 40:326–336. https://doi.org/10.1080/09670879409371908

Ploetz RC (2006) Fusarium wilt of banana is caused by several pathogens referred to as *Fusarium oxysporum* f. sp *cubense*. Phytopathology 96:653–656. https://doi.org/10.1094/phyto-96-0653

Ploetz RC (2015a) Fusarium wilt of banana. Phytopathology 105:1512–1521. https://doi.org/10.1094/phyto-04-15-0101-rvw

Ploetz RC (2015b) Management of fusarium wilt of banana: a review with special reference to tropical race 4. Crop Prot 73:7–15. https://doi.org/10.1016/j.cropro.2015.01.007

Ploetz RC, Pegg K (2000) Fusarium wilt in diseases of Banana, abaca and Enset. CABI Publishing, Wallingford, UK, pp 143–159

Ploetz RC, Nishijima GA, Rohrbach WT, Kg Ohr HD (1994) Compendium of tropical fruit diseases. Washington, DC. USA, The American Phytopathological Society, pp 7–28

Promusa (2021) Distribution Foc TR4. http://www.promusa.org/Tropical+race+4+-+TR4#Distribution. Accessed 16 March 2021

Rasiah V, Armour JD, Moody PW, Pattison AB, Lindsay S, Florentine S (2009) Characterizing and improving the deteriorating trends in soil physical quality under banana. Aust J Soil Res 47:574–584. https://doi.org/10.1071/sr08256

Rishbeth J (1957) Fusarium wilt of bananas in Jamaica: II. Some aspects of host-parasite relationships. Ann Bot 21:215–245. https://doi.org/10.1093/oxfordjournals.aob.a083561

Ritzinger CHSP, Fancelli M (2006) Integrated management of nematodes in the banana tree culture. Rev Bras Frutic 28:331–338. https://doi.org/10.1590/s0100-29452006000200041

Robinson JC, International CAB (1996) Bananas and plantains; CAB international: Wallingford. Conn, USA, p 312

Segura MR, Stoorvogel JJ, Garcia-Bastidas F, Salacinas-Niez M, Sandoval JA, Kema G (2015a) Soil management as an effective strategy for crop disease management: the case of Panama disease in banana. In proceedings of the book of abstract of the Wageningen soil conference: soil science in a changing world. Wageningen, The Netherlands, p 48

Segura RA, Serrano E, Pocasangre L, Acuna O, Bertsch F, Stoorvogel JJ, Sandoval JA (2015b) Chemical and microbiological interactions between soils and roots in commercial banana plantations (*Musa* AAA, cv. Cavendish). Sci Hortic 197:66–71. https://doi.org/10.1016/j.scienta.2015.10.028

Shabani F, Kumar L, Esmaeili A (2014) Future distributions of fusarium oxysporum f. spp. in European, middle eastern and north African agricultural regions under climate change. Agric. Ecosyst. Environment 197:96–105. https://doi.org/10.1016/j.agee.2014.08.005

Shen ZZ, Ruan YZ, Xue C, Zhong ST, Li R, Shen QR (2015a) Soils naturally suppressive to banana fusarium wilt disease harbor unique bacterial communities. Plant Soil 393:21–33. https://doi.org/10.1007/s11104-015-2474-9

Shen Z, Wang B, Lv N, Sun Y, Jiang X, Li R, Ruan Y, Shen Q (2015b) Effect of the combination of bio-organic fertiliser with bacillus amyloliquefaciens NJN-6 on the control of banana fusarium wilt disease, crop production and banana rhizosphere culturable microflora. Biocontrol. Sci Technol 25:716–731. https://doi.org/10.1080/09583157.2015.1010482

Shen Z, Penton CR, Lv N, Xue C, Yuan X, Ruan Y, Li R, Shen Q (2018) Banana fusarium wilt disease incidence is influenced by shifts of soil microbial communities under different monoculture spans. Microb Ecol 75:739–750. https://doi.org/10.1007/s00248-017-1052-5

Stover RH (1990) Fusarium wilt (Panama disease) of bananas—current status and control strategies. Acta Hortic 275:707–712. https://doi.org/10.17660/ActaHortic.1990.275.88

Summerell BA, Leslie JF, Liew ECY, Laurence MH, Bullock S, Petrovic T, Burgess LW (2011) Fusarium species associated with plants in Australia. Fungal Divers 46:1–27. https://doi.org/10.1007/s13225-010-0075-8

Taylor KE, Stouffer RJ, Meehl GA (2012) An overview of CMIP5 and the experiment design. B Am Meteorol Soc 93:485–498. https://doi.org/10.1175/BAMS-D-11-00094.1

Van Vuuren D, Edmonds J, Kainuma M et al (2011) The representative concentration pathways: an overview. Clim Chang 109(5):5. https://doi.org/10.1007/s10584-011-0148-z

Viloria JA, Olivares BO, García P, Paredes-Trejo F, Rosales A (2023) Mapping Projected Variations of Temperature and Precipitation Due to Climate Change in Venezuela. Hydrology 10(4):96. https://doi.org/10.3390/hydrology10040096

Waite BH, Dunlap VC (1953) Preliminary host range studies with *fusarium oxysporum* f. *cubense*. Plant Dis Rep 37:79–80

Waman AA, Bohra P, Sathyanarayana BN, Chandrashekar SC, Rani RT (2013) Are bananas (Musa spp.) really safe from their aesthetic relatives? Screening potential alternative hosts of *fusarium oxysporum* f. sp *cubense*. J Hortic Sci Biotechnol 88:559–562. https://doi.org/10.1080/14620316.2013.11513006

Wood-Sichra U, Joglekar AB, You L (2016) Spatial production allocation model (SPAM) 2005: Technical Documentation. Harvest Choice Working Paper., Harvest Choice: St. Paul, Minnesota, USA

World Climate Research Program (WCRP) (2021) WCRP Coupled Model Intercomparison Project (CMIP). https://www.wcrp-climate.org/wgcm-cmip. Accessed 20 Feb 2021

You L, Wood S, Wood-Sichra U, Wu W (2014) Generating global crop distribution maps: from census to grid. Agric Syst 127:53–60. https://doi.org/10.1016/j.agsy.2014.01.002

Zhang H, Mallik A, Zeng RS (2013) Control of Panama disease of banana by rotating and intercropping with Chinese chive (*Allium tuberosum* Rottler): role of plant volatiles. J Chem Ecol 39:243–252. https://doi.org/10.1007/s10886-013-0243-x

Zhong S, He YD, Zeng HC, Mo YW, Jin ZQ (2011) Effects of banana wilt disease on soil nematode community structure and diversity. Afr J Biotechnol 10:12759–12767. https://doi.org/10.5897/ajb11.1558

Chapter 4
Evaluation of the Incidence of Banana Wilt and its Relationship with Soil Properties

Abstract The incidence of "Cavendish" banana wilt (BW) caused by a fungal-bacterial complex is characterized by progressive wilting, rapid spread, and violent action, affecting the planted area and directly influencing banana production in the Aragua state of Venezuela. However, until now, there is no consensus on the soil characteristics associated with a high incidence of BW. The objective of this chapter was to identify soil properties potentially associated with BW incidence using supervised methods, for example, soil samples associated with banana plant lots in Venezuela. On those soils, sixteen soil physical and chemical variables were determined. An exploratory analysis with various machine learning algorithms and various resampling techniques was performed to select the best performing algorithm (see Chap. II). The random forest (RF) algorithm was used as a machine learning approach for classifying lots with high and low incidence of BW (see Chap. II). The analysis of the receiver operating characteristics (ROC) by RF revealed that the combination of Zn, Fe, Ca, K, Mn, and clay was able to accurately differentiate 84.1% of the banana lots with a sensitivity of 89.8% and a specificity of 72.4%. So far, our results open the field for further research in which we could quantitatively predict the risk of BW in banana fields based on available, or easy to gather, information, which in turn could allow farm managers to implement preventive measures to reduce BW risk.

Keywords Calcium · Clay · Iron · Machine learning · Random forest · Zinc

4.1 Introduction

Bananas (*Musa* spp.) represent an important crop for Venezuela's economy, predominantly based on oil. During the last 20 years, banana production has undergone slight reductions, reaching 650,051 tons in 2019, with a cultivated area of around 41,708 ha, partially due to the shortage of agricultural inputs (fertilizers and agrochemicals), problems with access to foreign currency to meet domestic demand,

B. O. Olivares Campos, *Banana Production in Venezuela*, The Latin American Studies Book Series, https://doi.org/10.1007/978-3-031-34475-6_4

and the inadequate management of agricultural policies and the impact of drought, pests, and diseases (FAO 2020).

Banana wilt (BW), also called "false Panama disease," was first described in South Africa by Deacon et al. (1985). Although some *Fusarium* species have been associated with BW plants, pathogenicity tests using those strains were not successful, and the etiology of BW could not be established. Both biotic and abiotic factors (Rey et al. 2016), including some physical and chemical soil characteristics and potentially pathogenic soil microorganisms (Domínguez et al. 2008), have been referred to as potential causes of BW. However, BW is a disease of an unknown etiology up to date and is mainly considered a physiological and metabolic plant disorder, which symptoms can be easily confounded with those of *Fusarium* wilt caused by *Fusarium oxysporum* f. sp. *cubense* (Foc), considered one of the most destructive diseases of bananas worldwide (Dita et al. 2018).

A differential diagnostic criterion has been established between *Foc* and BW based on the observation of symptoms and histological studies to determine abnormalities in phloem and xylem. Although it is not completely ruled out that some of the species studied could play a role in the development of the disease under abiotic conditions other than those tested, the results obtained point to an abiotic origin, in particular to certain soil characteristics that produce compaction, hypoxia, and blockage in the absorption of nutrients. Up to now, improving the structure, the texture, and the drainage of the soil, as well as the distribution of the irrigation dose in two or three applications per day, seems to be the only control tip that can be useful to farmers (Domínguez et al. 2008; Rey et al. 2020).

In the Aragua state of Venezuela, one of the main producing areas in the country, the yields of Cavendish bananas have been decreasing since 2006 associated with the BW disease (Martínez et al. 2020; Rey et al. 2020), increasing the concerns of farmers. However, since the causal agent of this disease has not been properly identified yet, its prevention and control are difficult.

The scientific literature and the evidence in the field of Martínez et al. (2020) and Rey et al. (2020) in Venezuela suggest that there is possibly a relationship between the properties of the soil that generate a stress situation in the plant caused by specific abiotic factors, which would enhance a deleterious effect of certain microorganisms such as fungi and bacteria (fungal-bacterial complex), inducing the expression of symptoms in the plant.

Thus, according to Rey et al. (2020), BW is associated with a fungal-bacterial complex and with some agroecological conditions characterized by silty soils presenting drainage problems and with nutritional imbalances, typical of lacustrine soils that are accentuated by unappropriated fertilization regimens in the last years. Additionally, the appearance and increase of the disease are associated with an average annual precipitation decrease and maximum temperature increase (Olivares et al. 2021).

Despite technological advances, it is difficult to find studies that relate soil properties to disease incidence through the use of supervised methods such as random forest (RF), orthogonal partial least squares discriminant analysis (OPLS-DA), and other algorithms. RF is a supervised learning classifier that can be used in complex situations (Hou et al. 2015; Ye et al. 2020) and has been proven to be a highly accurate classifier, but it has rarely been applied in the identification of soil properties associated with the incidence of diseases such as BW (Gomez-Selvaraj et al. 2020; Yuan et al. 2020).

In order to anticipate the potential occurrence of BW disease, it will be very valuable if certain soil characteristics can be associated with a major risk of occurrence of BW incidence. This chapter presents a study aimed to validate the hypothesis that it is possible to identify specific soil properties associated with a high incidence of BW using supervised methods such as RF and OPLS-DA, whose results can be of straightforward agronomic and environmental interpretation.

4.2 Results

4.2.1 Incidence of BW in Experimental Lots

The analysis of the identification of pathogenic microorganisms revealed the presence of bacteria (*Pectobacterium* and *Erwinia* genera) and fungi (*F. moniliforme*, *F. oxysporum*, and *F. solani*); these microorganisms have also been found by Sabadell (2003) in tissues with wilt symptoms from the Canary Islands (Spain) and recently by Rey et al. (2020) in lacustrine banana soils of Venezuela, but no vascular *Fusarium* isolates were recovered from the internal symptoms, which indicated that symptoms observed in the field plot were associated to BW and not to *Fusarium* wilt.

Symptoms of BW disease are shown in Fig. 4.1. Generally, yellowing begins on the lower or older leaves. The margin of each leaf turns pale green to yellow, necrotic stripes appear surrounded by a yellow margin, and the leaf eventually dies (Fig. 4.1a). The lower leaves die and hang from the pseudostem like a skirt (Fig. 4.1b). According to Beer et al. (2001), the base of the leaf remains green and healthy, while its distal part dies. Often one to four upper leaves remain green but are smaller in size, and their development stops. New leaf growth can occur, but the bunches in this case are generally small with short and thin bananas, which generates economic losses due to the rejection of the fruit in the market.

All the lots evaluated ($n = 78$) in the study area have BW disease. The percentage of lots with a low incidence (<1.90%) of BW reached 32.05% ($n = 29$), while the lots with high incidence (≥1.90%) represented 67.95% ($n = 49$) (Fig. 4.2). The highest incidence values were found in lot 36 with 8.47%, lot 32 with 5.97%, and lot 34 with 5.13% for the year 2017, while during 2016, the maximum incidence values were registered in lots 38 and 45 with 5.57% and 5.03%, respectively. On the other hand, lots 12, 13, and 17 presented low-incidence values that did not exceed 1.0% in both years of evaluation (Fig. 4.2).

Fig. 4.1 Symptoms of banana wilt disease in the study area. (**a**) The yellow margins on the leaves and the necrotic stripes are surrounded by the yellow margins on the lower or older leaves. (**b**) Set of dead leaves hanging from the pseudostem of a plant affected with banana wilt disease. (**c**) Longitudinal and ascending cracks, near the base of the pseudostem, with a wet appearance, with the presence of foul-smelling aqueous fluids that contribute to the weakening of the plant, and (**d**) consequent fall of the same, when it falls on the lesion advanced [Pictures by Juan C. Rey] (Maracay-Venezuela, 2021)

4.2.2 Description of Soil Properties in Experimental Lots

Figure 4.3 shows the result of the box plots and kernel density plots before and after the normalization of 16 variables. The figure provides an intuitive visualization of the data used. In general, the soils with a high incidence of BW presented loam to silty loam textures, with a predominance of particles with an equivalent diameter

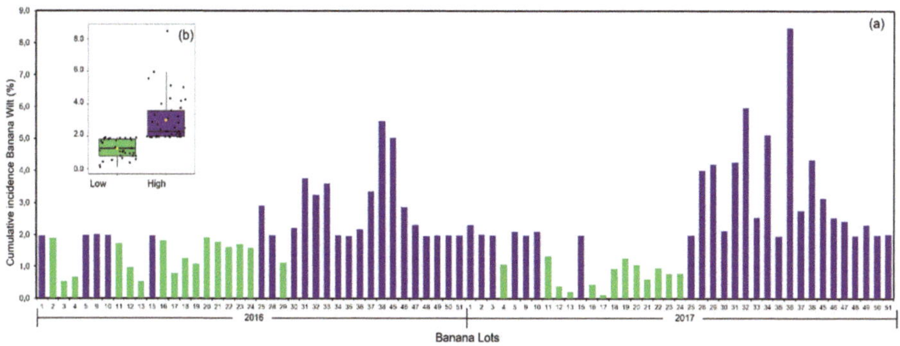

Fig. 4.2 (**a**) Cumulative incidence (%) of banana wilt in the study area during 2016 and 2017 (n = 78, mean = 2.17 ± 1.40%, min = 0.11%, max = 8.47%, asymmetry = 1.78; kurtosis = 4.46; P50 = 1.90%). (**b**) Box plots of banana lots with low (29) and high (49) incidence of BW

between 2 and 50 μm. In these soils, the banana lots classified as high incidence of BW showed high values of Na, Fe, and Mn and slightly higher pH values (Fig. 4.4).

On the other hand, the characteristics of the parental material of these soils produce very high levels of Ca. The limitations for the development of roots in these soils with a high incidence of BW could be associated with chemical conditions, such as the presence of high $CaCO_3$ content, the limiting ratios being Ca/Mg and Ca/K (*data not shown*). Sodium levels were high in most of the lots with a high incidence of BW, which could generate toxicity problems for plants and low structural stability in soils. Likewise, low levels of Cu were observed in banana lots with a low incidence of BW. The metabolic nature of these elements means that their deficiency can greatly affect the development of the crop. It is important to highlight that in some banana lots with a high incidence of BW, high levels of P were present on the surface, possibly due to overfertilization.

In very loamy soils, with low permeability and limited drainage and with nutrient imbalance, BW disease was more frequent. The differences in the incidence of the disease may be due to the fact that banana lots 32 to 38 have been planted for more than 10 years, while most of the lots 9 to 11 and a sector of banana lots 15 to 18 had been about 5 years old at the time this study was carried out. So, it is to be expected that in these last banana lots, the disease can advance rapidly. Likewise, it can be inferred that in soils with high incidence of BW, the clay content is slightly higher, whereas the K and Zn contents are slightly lower. However, a high incidence of BW occurs in those plant lots where the Ca content is higher, while the soils are more saline in depth.

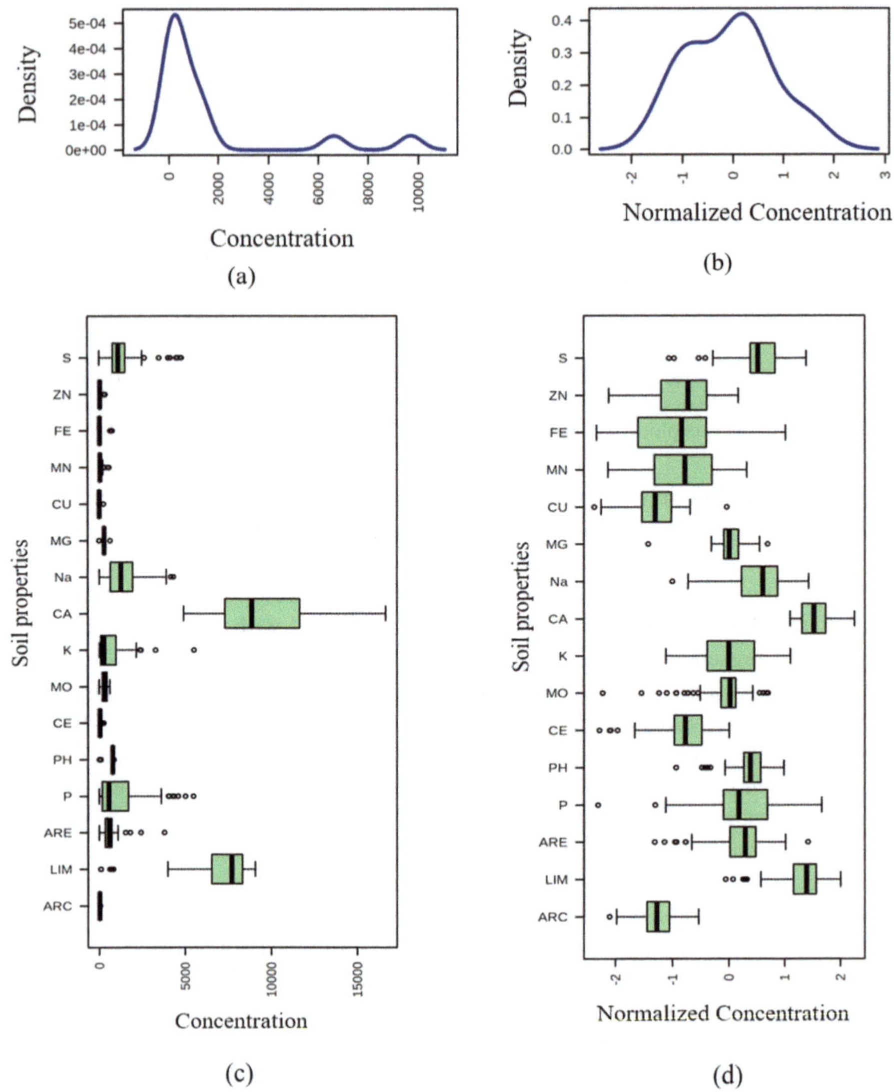

Fig. 4.3 Box plots and kernel density plots (**a**) before and (**b**) after normalization of 16 soil properties. The density plots are based on all samples and concentration of soil properties (**c**) before and (**d**) after normalization. The selected method for the data transformation was geometric mean and Log10 normalization

4.2.3 Exploratory Analysis

The area under the receiver operating characteristic (ROC) curve of linear discriminant analysis (Fig. 4.5a), RF (Fig. 4.5b), and radial basis function kernel SVM (Fig. 4.5f) were 0.80, 0.90, and 0.84, respectively, with synthetic minority oversampling technique (SMOTE), while the classification and regression tree algorithm

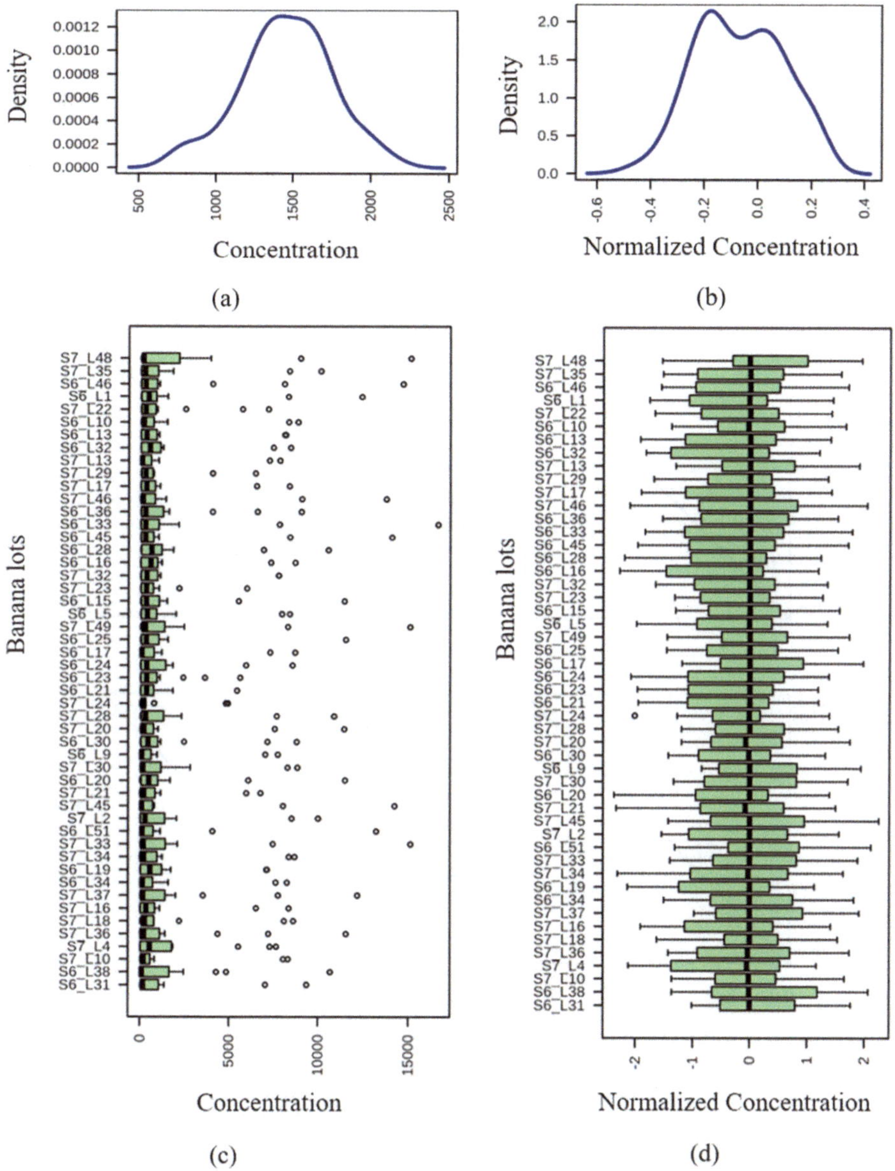

Fig. 4.4 Box plots and kernel density plots (**a**) before and (**b**) after normalization. The box plots show at most 50 cases of incidence of BW evaluated in the year 2016 (s6) and year 2017 (s7), due to space limits. The density plots are based on all samples. Concentration of soil properties of banana lots samples (**c**) before and (**d**) after normalization. The selected method for the data transformation was geometric mean and Log10 normalization

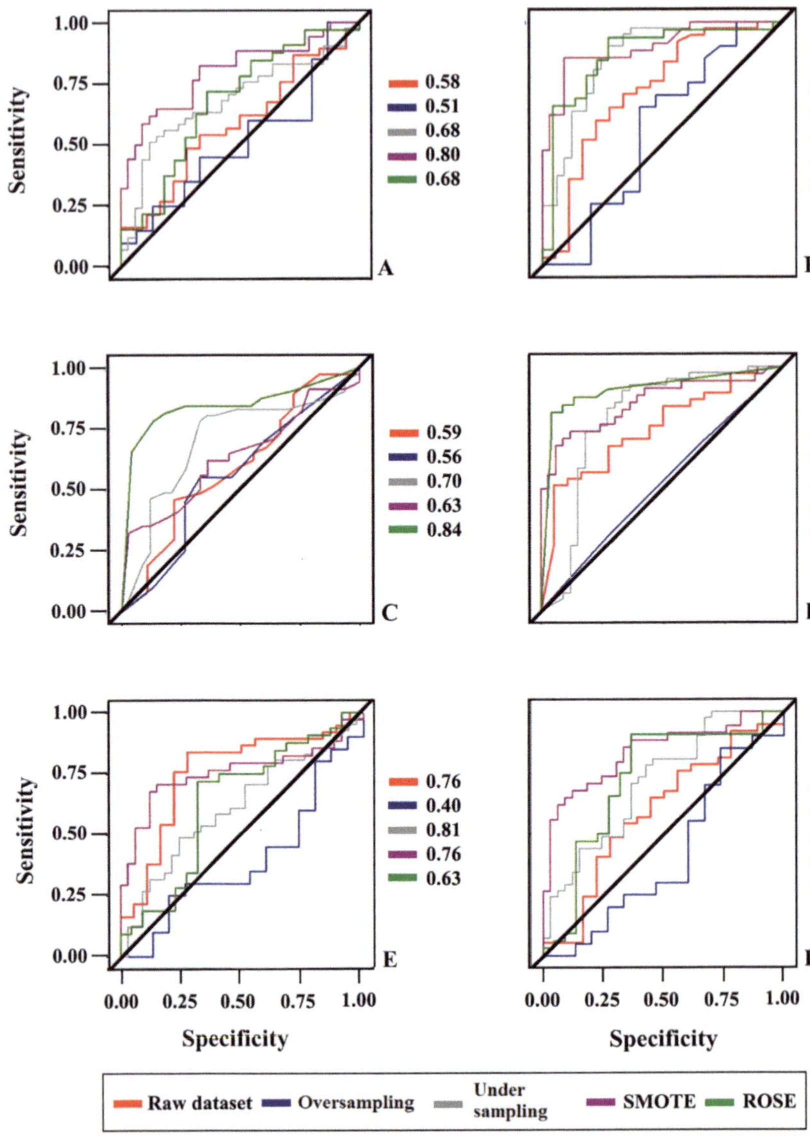

Fig. 4.5 Receiver operating characteristic (ROC) curves of linear discriminant analysis (**a**), random forest (**b**), classification and regression trees (**c**), C5.0 decision tree algorithm (**d**), linear kernel support vector machines (**e**), and radial basis function kernel-support vector machines (**f**) for sensitivity and specificity in predicting wilt banana diagnosis with the resampling methods evaluated

(Fig. 4.5c) and C5.0 decision tree algorithm (Fig. 4.5d) were 0.84 and 0.90, respectively, with random oversampling. On the other hand, the linear kernel SVM (Fig. 4.5e) showed an area under the ROC curve of 0.81 with undersampling

Table 4.1 Performance evaluation of different machine learning algorithms in the testing: linear discriminant analysis (LDA), classification and regression trees (CART), random forest (RF), C5.0 decision tree algorithm, linear kernel support vector machines (LSVM), and radial basis function kernel support vector machines (RSVM)

Model	Resampling	Sensitivity	Specificity	Accuracy	Accuracy IC 95%	Kappa
LDA	Raw dataset	0.00	0.82	0.61	(0.39–0.80)	−0.21
	Oversampling	0.67	0.82	0.78	(0.56–0.93)	0.47
	Undersampling	0.83	0.88	0.87	(0.66–0.97)	0.68
	SMOTE	1.00	0.71	0.78	(0.56–0.93)	0.56
	ROSE	0.33	0.82	0.70	(0.47–0.87)	0.17
CART	Raw dataset	0.50	0.82	0.74	(0.52–0.90)	0.32
	Oversampling	0.50	0.24	0.30	(0.13–0.53)	−0.17
	Undersampling	0.50	0.82	0.74	(0.52–0.90)	0.32
	SMOTE	1.00	0.65	0.74	(0.52–0.90)	0.49
	ROSE	1.00	0.18	0.39	(0.20–0.61)	0.10
RF	Raw dataset	0.50	0.88	0.78	(0.56–0.93)	0.40
	Oversampling	0.83	0.71	0.74	(0.52–0.90)	0.44
	Undersampling	1.00	1.00	1.00	(0.85–1.00)	1.00
	SMOTE	1.00	0.88	0.91	(0.72–0.99)	0.80
	ROSE	0.50	0.65	0.61	(0.39–0.80)	0.13
C5.0	Raw dataset	0.50	0.82	0.74	(0.52–0.90)	0.32
	Oversampling	0.50	0.41	0.43	(0.23–0.66)	−0.06
	Undersampling	0.83	1.00	0.96	(0.78–1.00)	0.88
	SMOTE	1.00	0.76	0.83	(0.61–0.95)	0.63
	ROSE	0.67	0.65	0.65	(0.43–0.84)	0.26
LSVM	Raw dataset	0.33	0.53	0.48	(0.27–0.69)	−0.11
	Oversampling	0.50	0.88	0.78	(0.56–0.93)	0.40
	Undersampling	0.67	0.88	0.83	(0.61–0.95)	0.55
	SMOTE	1.00	0.71	0.78	(0.56–0.93)	0.56
	ROSE	0.33	0.82	0.70	(0.47–0.87)	0.17
RSVM	Raw dataset	0.00	1.00	0.74	(0.52–0.90)	0.00
	Oversampling	0.00	1.00	0.74	(0.52–0.90)	0.00
	Undersampling	0.50	0.88	0.78	(0.56–0.93)	0.40
	SMOTE	0.83	0.82	0.83	(0.61–0.95)	0.59
	ROSE	0.00	1.00	0.74	(0.52–0.90)	0.00

technique. In general, the raw dataset presented low values for the fitted models, indicating that it is necessary to consider the resampling approach for our dataset.

In our study (Table 4.1), random forest was significantly more accurate than C5.0 decision tree algorithm and linear discriminant analysis, with values of 1.0, 0.96, and 0.87, respectively, through the undersampling technique; radial basis function kernel support vector machines performed better than linear kernel support vector machines and the classification and regression tree algorithms. There was no significant difference between linear and radial SVM with SMOTE. Although results were not significantly different, ease of model construction was much greater for random forest than for support vector machines. In random forest, only one key parameter (number of trees) is adjusted; support vector machine models need to adjust at least 4–5 parameters. Additionally, the meaning of some parameters is unfamiliar to farmers. Considering ease of model construction, random forest is a better model for use in diagnosing wilt on banana.

In general, a similar behavior was observed in that the undersampling technique was the one that presented a greater area under the ROC curve for most of the evaluated algorithms. In addition, it was observed in Table 4.1 that the ROSE resampling technique showed a lower performance for the dataset and for the evaluated algorithms, standing out for presenting, in general, low values of the area under the ROC curve, both for the sets of validation such as testing, low precision, sensitivity, and specificity.

The above results confirmed the advantage of the proposed RF method over the C5.0, LDA, LSVM, RSVM, and CART methods. As it can be seen from the Tables 4.1 and 4.2, the RF method gives the best results when the four statistical indices are considered, which means that all the instances or lots of bananas with incidence are correctly classified. The number of trees and the random characteristics affect the performance of the random forest classifier. The classification performance of the linear SVM and radial kernel function in Table 4.1 shows that it cannot be an applicable method for the identification of lots with incidence of banana wilt. The C5.0 decision tree achieves remarkable performance in both cases: test dataset and raw dataset. C5.0 misclassified few instances belonging to high-incidence lots. This implies that C5.0 may also be suitable for classifying banana lots based on soil characteristics (Table 4.2).

4.2.4 Formal Analysis with Wilcoxon Rank Test

For direct comparison of the levels of the soil variables, the Wilcoxon analysis was used to identify critical variables between the groups with low and high incidences of BW. The analysis revealed a total of six significant soil variables (adjusted p-value <0.05) (Table 4.3): Zn, Ca, Fe, clay, Mn, and K. In our study, a small fraction of false positives could be accepted to substantially increase the total number of discoveries; therefore the false discovery rate (FDR) obtained is usually appropriate and useful. The FDR is the rate at which the so-called significant features are actually null. An FDR of 5% means that, among all the characteristics called significant, 5% of these are actually null. For example, if a study publishes statistically significant results for an FDR of 5.0%, the reader is confident that at most 5.0% of the results deemed significant are actually false positives. The significant and most important soil variables that were responsible for the observed differentiation between the two BW incidence groups are shown in Fig. 4.6.

Table 4.2 Results of sensitivity, specificity, accuracy, the 95% confidence interval, and kappa statistic for the classifying algorithms using the kappa criterion (kappa >0.60) in the original and test databases

Algorithm[a]	Resampling Method[b]	Sensitivity		Specificity		Accuracy		Accuracy IC 95%		Kappa	
		TD	RD	TD	RD	TD	RD	TD	RD	TD	RD
RF	Undersampling	1.00	1.00	1.00	0.94	1.00	0.96	(0.85–1.00)	(0.89–0.99)	1.00	0.91
C5.0	Undersampling	0.83	0.96	1.00	0.98	0.96	0.97	(0.78–1.00)	(0.91–1.00)	0.88	0.94
RF	SMOTE	1.00	0.92	0.88	0.80	0.91	0.83	(0.72–0.99)	(0.73–0.91)	0.80	0.65
LDA	Undersampling	0.83	0.83	0.88	0.76	0.87	0.78	(0.66–0.97)	(0.67–0.87)	0.68	0.54
C5.0	SMOTE	1.00	0.88	0.76	0.74	0.83	0.78	(0.61–0.95)	(0.67–0.87)	0.63	0.55

[a]Algorithms with kappa >0.60, such as RF, random forest; C5.0, classification trees with algorithm C5.0; and LDA, linear discriminant analysis
[b]Resampling methods. *TD* test dataset, *RD* raw dataset

Table 4.3 Important variables selected by the Wilcoxon rank test with a threshold 0.05. (Olivares et al. 2022)

Variable	V	*p*-value	-log10(p)	FDR
Zn	1199	9.52E-05	80.215	1.52E-03
Ca	222	2.46E-02	56.086	1.38E-01
Fe	223	2.60E-02	55.858	1.38E-01
Clay	357	0.001	29.618	0.004
Mn	386	0.003	25.052	0.009
K	921	0.005	22.414	0.015

Note: V: V-statistics. These values are based on the unpaired data. FDR: false discovery rate

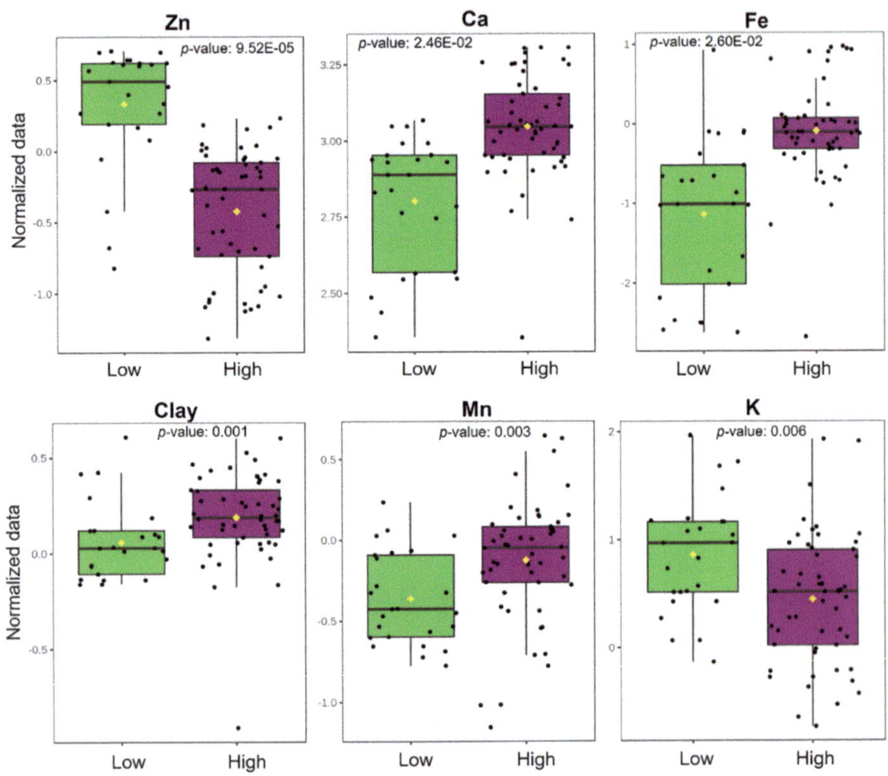

Fig. 4.6 Box plots of levels of significant soil variables based on the Wilcoxon rank test. (Olivares et al. 2022)

4.2.5 *Identification of Important Soil Variables*

The variable importance in the projection (VIP) values was obtained from the OPLS-DA model. The VIP was taken for selection, and those variables with VIP > 1 were considered as possible candidate variables for group discrimination (Fig. 4.7).

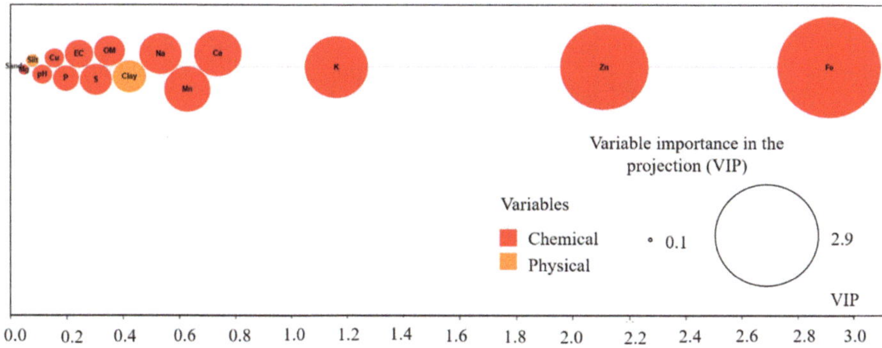

Fig. 4.7 The variable importance in the projection (VIP) values was obtained from the OPLS-DA model

Accordingly, the analysis revealed prominent values in three variables: K, Fe, and Zn. On the other hand, as shown in Fig. 4.8a, the OPLS-DA allowed us to analyze the information collected in the predictive component independently of the orthogonal components. That is, it allowed us to separate the variability responsible for the discrimination from the noise generated by the uncorrelated variability. For this reason, the OPLS-DA was the method chosen for the selection of variables considered relevant in the discrimination of groups. Also, based on the loading values >0.2, OPLS-DA identified six critical variables: clay, Mn, K, Ca, Fe, and Zn (Fig. 4.8b). Besides, this OPLS-DA model showed a proper fitting of the data ($R^2Y = 0.61$, p-value <0.01) and exhibited good predictive power ($Q^2 = 0.50$, p-value <0.01) (Fig. 4.8c).

4.2.6 *Classifier Performance and Accuracy Assessment*

Table 4.4 shows the lower limit, upper limit, mean, and standard deviation of soil properties in banana lots with low and high incidences of BW. Fig. 4.9 shows the measures of the importance of the soil variables selected by the RF model. The results establish the frequency that an independent variable is selected greater than/equal to a defined significance threshold (0.5). The mean decrease accuracy (MDA) allows us to visualize the relative impact on the performance of the RF classifier by subtracting each specific soil variable (Fig. 4.9).

Figure 4.10 shows the classification results after RF analysis; the receiver operating characteristic (ROC) curve of the best-performing model indicated an area under the ROC curve (AUC) of 0.91 (95% confidence interval CI, 0.80% to 0.99%) (Fig. 4.10a). The scores plot (Fig. 4.10b) shows the predicted class probabilities for all samples included in the analysis, indicating the correct classification of 44 banana lots out of 49 with a high incidence of BW and 21 banana lots out of 29 with a low incidence.

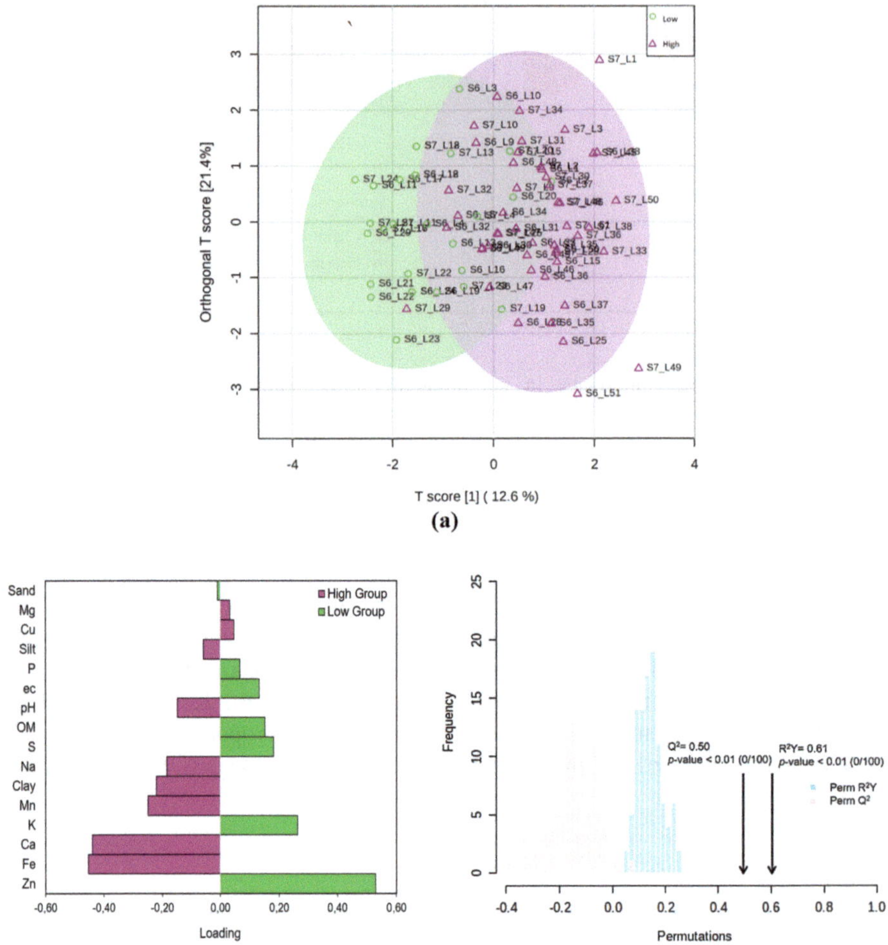

Fig. 4.8 (**a**) OPLS-DA score plot of all soil variables, based on separation of the groups (low incidence of BW group, $n = 29$; high incidence of BW group, $n = 49$). (**b**) Loading plot weights of each variable selected from OPLS-DA; the color indicates the class in which the variable has the maximum level of expression. (**c**) Internal validation of the corresponding OPLS-DA model by permutation analysis ($n = 100$); the fraction of the variance of descriptor class response (Y) (R^2Y) = 0.61 (blue bars), p-value <0.01; the fraction of the variance predicted (cross validated) (Q^2) = 0.50 (red bars), p-value <0.01. (Olivares et al. 2022)

The results showed the great power of the RF classifier to correctly differentiate lots of bananas with high or low BW incidence. Furthermore, our proposed system reached 89.80% of sensitivity and 72.40% specificity in the test dataset, which implies that most banana lots with low BW incidence were correctly classified with a false-negative (FN) rate of 5/49, and most of the banana lots with high BW incidence were also correctly classified with a false-positive (FP) rate of 8/29 (Fig. 4.10c).

Table 4.4 Lower limit, upper limit, mean, and standard deviation (SD) of soil properties in banana lots with the low and high incidences of BW

	Low-incidence group (n = 29)			High-incidence group (n = 49)		
Variables	Low LL	Mean ± SD	Low UL	High LL	Mean ± SD	High UL
Zn (mg/kg)	1.60	28.90 ± 11.28	36.80	0.36	5.81 ± 5.32	30.40
Fe (mg/kg)	0.04	4.43 ± 12.64	64.00	0.06	17.62 ± 25.0	78.40
Ca (mg/kg)	4936.0	6848.00 ± 1070.69	8920.0	6472.0	11051.92 ± 2597.26	16,448.0
Clay (%)	5.00	14.52 ± 8.52	31.00	1.00	16.85 ± 7.49	40.00
K (mg/kg)	5.60	156.04 ± 211.79	984.00	1.50	89.12 ± 235.94	1336.0
Mn (mg/kg)	1.60	6.18 ± 6.51	33.60	0.80	11.02 ± 12.05	58.40

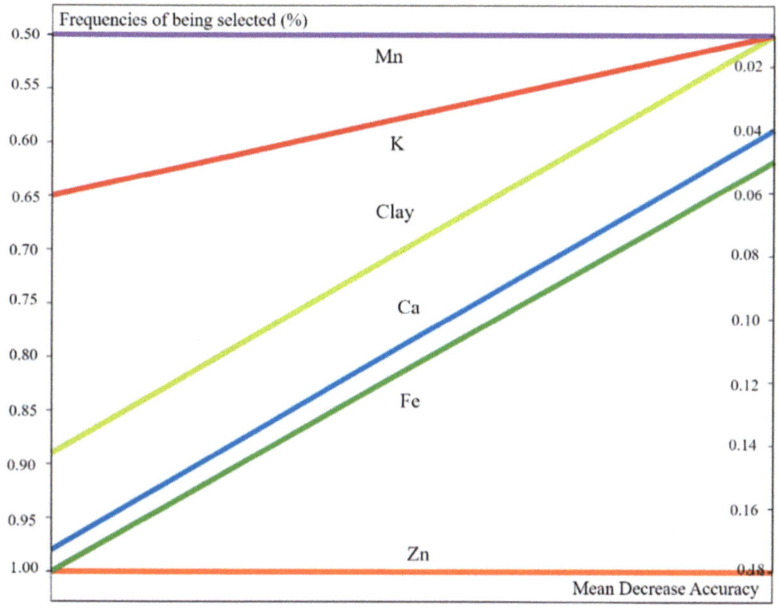

Fig. 4.9 Frequencies of variables being selected (%) and mean decrease accuracy of the model soil variables with random forest (accuracy, 84.10%)

4.3 Discussion

4.3.1 Global Distribution of BW Incidence

Despite the existing literature on banana wilt, its etiology is still unknown, due to the few studies that link biotic and abiotic factors with the incidence of BW. Particularly, these studies are represented in Costa Rica whose incidence of BW was 7.3%. (Lichtemberg et al. 2010); in Colombia, where an incidence of 0.3% has been reported in some banana-producing areas with a prevalence of 4.3% (Merchán 2002); and finally in banana areas corresponding to 15 provinces of

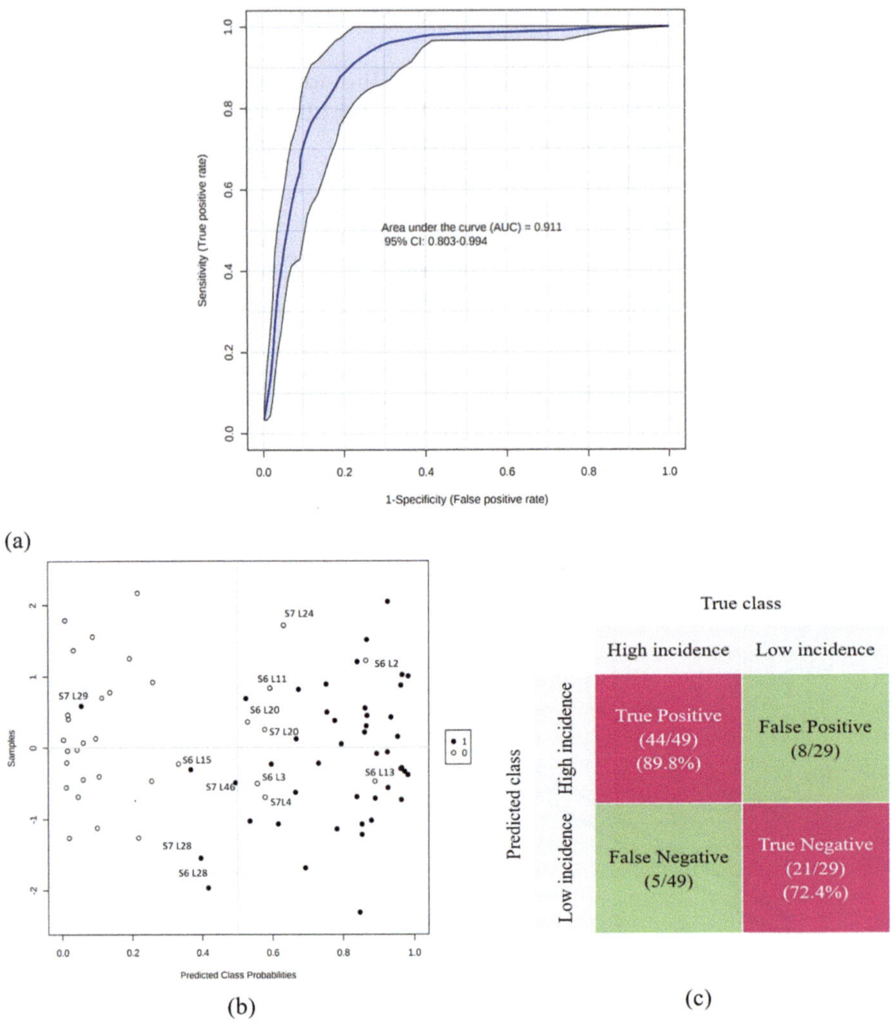

Fig. 4.10 Classification of banana lots according to the incidence of banana wilt (BW). (**a**) Receiver operating characteristic (ROC) curve after obtained by Random Forest as classification method. The values generated for the area under the ROC curve (AUC) (0.91) along with the 95% confidence intervals (CI) (0.80–0.99) are given within the graph and accuracy: 84.10%. (**b**) Predicted class probabilities for each banana lot, allowing display of misclassified banana lots (lots of high BW incidence are shown as black dots; lots of low BW are shown as white dots). Since a balanced subsampling approach is used for model training, the classification limit is always in the center ($x = 0.5$, the dotted line). (**c**) Confusion matrix showing the number of true positives (44/49), true negatives (21/29), false positives (8/29), and false negatives (5/49). Sensitivity and specificity are given in the regions highlighted in purple, being 89.80% and 72.40%, respectively. (Olivares et al. 2022)

Indonesia, whose average incidence of BW reached values of 24.0% (Hermanto et al. 2011). The report by Kangire and Rutherford (2001) indicates that in some villages in western Uganda, more than 70.0% of banana farms were affected with up to 25.0% incidence of BW, the symptoms of which were associated with local and introduced bananas.

A contribution of special relevance in the commercial farm evaluated in this study is represented by the incidence values of BW reported by the findings of Martínez et al. (2016), Ramírez et al. (2016), and Rey et al. (2020) whose results establish that the BW incidence ranges from 0.32% to 11.41% considering the variability of the banana lots. The findings of our study coincide with what was reported by Martínez et al. (2016), Ramírez et al. (2016), and Rey et al. (2020), with emphasis on the appearance of the largest number of plants with symptoms located in the area of lots 31 to 46 of the entire sampled area. Therefore, a relationship could be established between the physical and chemical properties of the soil and some aspects of poor management of banana plantations, particularly inadequate fertilization and the absence of soil conservation practices that is expressed in a notable nutritional imbalance in these banana soils.

4.3.2 Causal Soil Properties

There was an old and widely held belief in different banana areas of the world that this BW disease was *Fusarium* wilt. Unfortunately, very little research has been done regarding this disease in Venezuela, and, possibly due to initial perceptions, this did not help confirm or refute the belief or, more importantly, assist in its management or control.

Preliminary research in Uganda by Kangire (1998) established that soils on farms affected by BW generally showed higher calcium, potassium, and phosphorus contents in the vicinity of symptomatic plants, compared to areas where BW was not apparent. It was also determined that the levels of Ca, K, and P increased in soils managed with organic waste, green manures, and manure. These findings were the starting point for the development of the next studies by Beer et al. (2001), Sabadell (2003), Martínez et al. (2020), and Rey et al. (2020), which ratified the hypothesis that the physical and chemical properties of the soil are closely linked to the occurrence of BW.

4.3.2.1 Soil Chemical Properties

In the case of the commercial farm located in Aragua, Venezuela, it was determined that the soil differences of the properties such as Zn, Fe, Ca, K, Mn, and clay in the areas of high and low incidence of BW have a determining weight in the expression of the disease in the field. K contents are highest (5.0–984 mg/kg) in the group of lots with a low incidence of BW. However, Ca contents are excessively high in both groups, more notably the concentrations in the lots of high incidences of BW (6472–16,448 mg/kg), mainly because these soils are of lacustrine origin, this being a key characteristic to understand the deficiencies of Mg and K in banana soils (López and Espinoza 1995).

It is convenient to indicate that the high calcium content found in these soils does not necessarily coincide with the Ca content in the plant, that is, it is not the same real amount of calcium absorbed by the plant, particularly in the dry season (December to April) in Aragua, when soil moisture decreases and Ca movement in the profile is achieved (Turner et al. 2007). Also, the good assimilation of these nutrients is based on the equilibrium relationships between Ca, Mg, and K. According to what was reported by Díaz et al. (2007), these nutrients keep a close analogy with the exchangeable Ca content in the soil.

The research by Moreno et al. (1999) showed that plants susceptible to diseases presented low yields mainly associated with low K availability due to deficiencies or imbalances in the K/Ca/Mg ratios; otherwise, the excessive applications of K caused an excessive accumulation of this nutrient in the soil, also inducing imbalances between K/Ca/Mg. Therefore, maintaining an adequate balance in the content of K/Ca/Mg guarantees an availability, absorption, and functionality of these, as well as a positive effect on the production and quality of the banana fruit, and consequently that the plant is not susceptible to diseases such as BW.

The microelements such as Fe (0.1–78 mg/kg) and Mn (0.8–58 mg/kg) are present at high levels in the group of lots with a high incidence of BW, while Zn is at low levels (0.3–30 mg/kg) (Table 4.4). According to what was reported by Gutiérrez-Boem (2016), the high Fe and Mn contents could be related to the higher clay content in soils whose property could cause poor drainage problems in the soil and consequently an increase in the solubility of Fe^{2+} and Mn^{2+} due to excessive moisture conditions.

An investigation by Borges Pérez et al. (1991) on banana soils in the Canary Islands (Spain) determined that the application of Zn to the soil in areas with notable Zn deficiencies significantly reduces the incidence and severity of BW. Therefore, the low levels of Zn in banana lots with a high incidence of BW could favor the appearance of BW.

4.3.2.2 Soil Physical Properties

Generally, the spatial distribution of 2:1 type of clay predominates in the evaluated banana farm; therefore, the high cation exchange capacity and high Ca and K contents could have an influence on the ionic imbalances in this type of soil. Likewise, the water table in the Lake Valencia area, together with the compaction of the soil in several banana lots of the farms due to inadequate practices, limits the normal growth of the roots in the rainy season (Díaz et al. 2007).

The results of Serrano (2003) indicate that the areas with the highest vermiculite content and high percentage of micropores directly affect the availability of K. On the other hand, when there is a limitation in the circulation of oxygen in the soil, certain fermentation processes increase that cause the death of the tips of banana plants (Aguilar et al. 2003; Turner et al. 2007).

The findings of Domínguez (2008) in banana soils of the Canary Islands showed a notorious tendency to the formation of stable clay aggregates; coupled with excess irrigation, it generated anaerobiosis in these soils and therefore high Fe content, which caused soil compaction in the dry state and a greater manifestation of BW symptoms in those areas.

These relationships of clay content (1–40%) with water and the detrimental effect of compaction in banana soils result in a decrease in productivity, plant height, and a reduction in the number of children plants in the banana production unit. Additionally, according to the results of Dorel (1993) and Sabadell (2003), the most significant effect would be related to the reduction of the absorption of N, P, K, and Ca and the massive absorption of Mn.

The results of our analysis established that the heavy texture in the lots with high incidence of BW favors the appearance of symptoms, agreeing with other studies that have found that this disease develops in the presence of soils with high humidity and heavy texture (Deltour et al. 2017) and poor drainage (Lahav and Israeli 2019), favoring infection by deleterious microorganisms in the lateral rootlets.

The research by Rey et al. (2020) showed high correlations of variables such as organic carbon, sand and silt content, exchangeable Mg, and the Ca/Mg ratio with the incidence of BW. The soils evaluated are of lacustrine origin and very silty, with low permeability and limited drainage and with nutrient imbalance; therefore, it is presumed that these characteristics exert a direct influence on a high incidence of BW. Also, the reported finding establishes that the most important properties in the differentiation of banana lots with different levels of BW incidence were the C/N ratio, the content of K and Zn, and the relationships between the exchangeable cations Ca, Mg, and K, this finding being similar to the results of this study.

The spatial variability of the BW incidence was wide throughout the evaluated farm, and the areas with high incidence were located on very silty soils with certain drainage problems, nutrient deficiencies, and nutritional imbalance, according to Rey et al. (2020), due to the lacustrine origin and the lack of cycles of suitable and optimal fertilization in recent years.

4.3.3 Current Research with Machine Learning Algorithms

Currently, the use of modern techniques represented by machine learning and deep learning algorithms to identify characteristics of banana agroecosystems is becoming more evident. The scientific literature reports various investigations in the field of automatic learning for the detection and diagnosis of banana diseases, implementing RF (Owomugisha et al. 2014; Ma et al. 2017; Sangeetha et al. 2020; Ye et al. 2020; Gomez-Selvaraj et al. 2020), decision trees (Owomugisha et al. 2014), support vector machine (SVM) (Vipinadas and Thamizharasi 2016; Hou et al. 2015; Aruraj et al. 2019; Ye et al. 2020), and artificial neural networks (Ye et al. 2020), among others. Indeed, the exposed results establish that RF was able to identify the soil properties that could favor the development of the BW disease, serving as a fundamental base to investigate the etiology and the role of the edaphic factors that contribute to the development of BW, as well as the means to identify the mechanisms for its effective management based on timely fertilization.

In the scientific literature, certain machine learning analysis approaches have been used to detect diseases such as *Fusarium* wilt and black sigatoka using aerial imagery, but none of these studies have used in situ soil property data to classify the occurrence of a disease like BW. This aspect makes the study of BW in banana soils in Aragua a pioneer in identifying soil properties with supervised methods such as OPLS-DA and the RF algorithm, resulting in useful information regarding the application of these innovative techniques based on algorithms. The sores shown indicate that Zn, Fe, Ca, K, Mn, and clay could be promising soil indicators to classify banana lots susceptible to a higher incidence of BW disease in lacustrine soils.

Of all the classifiers used in previous work by Sangeetha et al. (2020), Ye et al. (2020), and Gomez-Selvaraj et al. (2020), the RF algorithm presented an important advantage in disease classification performance. The characteristics of the RF classifier and the way in which the most important soil variables are selected through the OPLS-DA determine the performance of the RF classifier. However, the conclusions of the studies by Gomez-Selvaraj et al. (2020) and Ye et al. (2020) state that the quality and representativeness of the information obtained, the performance of the feature extraction algorithm, and the subsets used for training and testing purposes can affect the accuracy of classifying banana lots with different levels of BW incidence. In summary, the results shown here point to RF as the algorithm with the best performance in the differentiation of banana lots with high or low incidence of BW, which means that the model reported here provides an easy, fast, and economical method to classify with precision the risk of incidence of BW in bananas from tropical territories.

4.4 Conclusion

The efficiency of the fertilizer will depend on the balance between cations, such as potassium (K^+), calcium (Ca^{2+}), magnesium (Mag^2), and sodium (Na^+), and anions, such as chloride (Cl) and nitrate (NO_3). Potassium, for example, favors the transport of soil nutrients, while calcium and zinc help control BW. On the other hand, if the plant does not acquire enough sulfur, glutathione will not be produced, a substance that defends it against diseases. The banana plant morphologically has an extractive root system and presents rapid vegetative growth, which causes a great capacity for extracting nutrients from the soil. Fertilization must be done in a granular, organic, and foliar way depending on the needs of the crop in relation to the appropriate foliar and soil analyzes.

In the present study, by using a random forest analysis approach, we identified that the risk of the low or high incidence of BW in a banana farm in Venezuela could be associated with differences in six key soil variables including Zn, Fe, K, Ca, Mn, and clay content. The findings may contribute to increasing our understanding of the basic mechanisms and progression of BW incidence and indicate that these soil variables are potentially determining factors of a risk of high BW incidence in tropical lacustrine soils of Venezuela.

Although the RF analysis performed well in this particular study and its performance in other banana areas in Venezuela has not been proven yet, we consider that this machine learning algorithm using soil properties as indicators has the potential to be further explored as a simple and effective tool in banana areas with the risk of developing BW.

Obtaining relevant information from banana soil data could not only help improve the farmer's experience but also help make relevant decisions from the early stages of the appearance of symptoms in the field. In this sense, the algorithms evaluated here would allow analyzing, interpreting, and predicting future scenarios that help prevent and reduce the burden of diseases in Venezuelan banana lots and their impact on production.

References

Aguilar EA, Turner DW, Gibbs DJ, Armstrong W, Sivasithamparam K (2003) Oxygen distribution and movement, respiration and nutrient loading in banana roots (*Musa spp.* L.) subjected to aerated and oxygen-depleted environments. Plant Soil 253(1):91–102

Aruraj A, Alex A, Subathra MSP, Sairamya NJ, George ST, Ewards SEV (2019) Detection and classification of diseases of Banana plant using local binary pattern and support vector machine. 2019. 2nd international conference on signal processing and communication (ICSPC). pp. 231–235. https://doi.org/10.1109/ICSPC46172.2019.8976582

Beer Z, Hernández J, Sabadell S (2001) Enfermedad del falso mal de Panamá en banano. Enfermedades de musa: Hoja divulgativa N° 9. INIBAP, Montpellier, Francia. 4 p. https://bit.ly/386KkUS. Accessed 07 February 2021

Borges-Pérez AF, Bravo JJ, Perez JF, Lopez I (1991) Enhanced resistance of banana plants (*dwarf Cavendish*) to *fusarium oxysporum* f. sp. *cubense* by controlled Zn nutrition under field conditions. Banan Newsl 14:24–26

Deacon JW, Herbert JA, Dames J (1985) False Panama disorder of bananas. Citrus and Subtropical Fruit Research Institute Information Bulletin, In, pp 15–18

Deltour P, Franca SC, Pereira OL, Cardoso I, De Neve S, Debode J, Hofte M (2017) Disease suppressiveness to fusarium wilt of banana in an agroforestry system: influence of soil characteristics and plant community. Agric Ecosyst Environ 239:173–181. https://doi.org/10.1016/j.agee.2017.01.018

Díaz A, Cayón G, Mira JJ (2007) Calcium metabolism and its relationship with "maturity bronzing" in banana fruits. A review Agronomía Colombiana 25(2):280–287

Dita M, Barquero M, Heck D, Mizubuti ESG, Staver CP (2018) Fusarium wilt of banana: current knowledge on epidemiology and research needs toward sustainable disease management. Front Plant Sci 9:1468. https://doi.org/10.3389/fpls.2018.01468

Domínguez J, Negrin MA, Rodriguez CM (2008) Soil potassium indices and clay-sized particles affecting banana-wilt expression caused by soil fungus in banana plantation development on transported volcanic soils. Commun Soil Sci Plant Anal 39(3–4):397–412. https://doi.org/10.1080/00103620701826522

Dorel M (1993) Banana development in an andosol in Guadeloupe: effect of soil compaction. Fruits 48:83–88. https://n9.cl/vahwkz

FAO (2020) FAOSTAT Database: Area, production and export of bananas and plantains https://bit.ly/34z8QN8. Accessed 5 Mar 2020

Gomez-Selvaraj M, Vergara A, Montenegro F, Alonso Ruiz H, Safari N, Raymaekers D, Ocimati W, Ntamwira J, Tits L, Omondi AB, Blomme G (2020) Detection of banana plants and their major diseases through aerial images and machine learning methods: a case study in DR Congo and Republic of Benin. ISPRS 169:110–124. https://doi.org/10.1016/j.isprsjprs.2020.08.025

Gutiérrez-Boem FH (2016) Micronutrientes. In Carina R. Álvarez C.R and Rimski-Korsakov, H. (Eds) Manejo de la fertilidad del suelo en planteos orgánicos, pp. 106–113. Editorial Facultad de Agronomía, Universidad de Buenos Aires, Argentina

Hermanto C, Sutanto A, Jumjunidang EHS, Daniells JW, O'Neill WT, Sinohin VGO, Molina AB, Taylor P (2011) Incidence and distribution of fusarium wilt disease of banana in Indonesia. Acta Hortic 897:313–322. https://doi.org/10.17660/ActaHortic.2011.897.43

Hou JC, Hu YH, Hou LX, Guo KQ, Satake T (2015) Classification of ripening stages of bananas based on support vector machine. Int J Agric Biol 8(6):99–103. https://doi.org/10.3965/j.ijabe.20150806.1275

Kangire A (1998) Fusarium wilt (Panama disease) of exotic bananas and wilt of east African highland bananas (*Musa*, AAA-EA) in Uganda. Department of Agriculture, Reading University, Thesis submitted for the degree of Doctor of Philosophy, p 304

Kangire A, Rutherford M (2001) Desorden parecido al marchitamiento en los bananos de Uganda. Enfermedades de Musa: Hoja divulgativa 10:1–4

Lahav E, Israeli Y (2019) Mineral deficiencies of banana. In: Jones D (ed) Diseases of banana, Abacá and Enset. CAB International Publishing, Wallingford, Oxon, UK, pp 475–486

Lichtemberg PS, Pocasangre LE, Sikora RA, Staver C (2010) Current status of Fusarium Wilt (*Fusarium oxysporum* f. sp. *cubense*) in bananas Gros Michel (AAA) at smallholder level of Costa Rica. XIX Reunión Internacional ACORBAT 2010. Medellín. Colombia, pp 32–44

López A, Espinoza J (1995) Manual de nutrición y fertilización de banano. International Plant Nutrition Institute, Quito, Ecuador, p 86

Ma L, Fu T, Blaschke T, Li M, Tiede D, Zhou Z, Ma X, Chen D (2017) Evaluation of feature selection methods for object-based land cover mapping of unmanned aerial vehicle imagery using random Forest and support vector machine classifiers. ISPRS Int J Geo Inf 6(51). https://doi.org/10.3390/ijgi6020051

Martínez G, Rey JC, Castro L, Micale E, López O, Pargas R, Manzanilla E (2016) Cavendish banana wilt in the central region of Venezuela associated with a fungus-bacteria complex. In: Resúmenes de Ponencias, pp 34, ACORBAT Meeting, 2016, Miami USA

Martínez GE, Rey-Brina JC, Rodríguez D, Jiménez C, Rodríguez Y, Rumbos R, Pargas-Pichardo R, Martínez E (2020) Análisis de la situación fitopatológica actual de las musáceas comestibles en Venezuela. Agronomía Tropical 70:1–20. https://doi.org/10.5281/zenodo.4323273

Merchán VM (2002) Manejo Integrado de plagas del plátano y banano. In: Acorbat. Memorias XV Reunión, Asociación de Bananeros de Colombia AUGURA. Medellín Colombia, pp 354–361p

Moreno M, Fernández L, Sosa L, Nava JC (1999) Effect of potassium, calcium and magnesium on banana tree yield (*Musa* AAA, Cavendish subgroup, great dwarf clone). Rev Fac Agron (LUZ) 16(Supl 1):114–123

Olivares B, Paredes F, Rey JC, Lobo D, Galvis-Causil S (2021) The relationship between the normalized difference vegetation index, rainfall, and potential evapotranspiration in a banana plantation of Venezuela. STJSSA 18:58–64. https://doi.org/10.20961/stjssa.v18i1.50379

Olivares B, Vega A, Calderón MAR, Rey JC, Lobo D, Gómez JA, Landa BB (2022) Identification of soil properties associated with the incidence of Banana wilt using supervised methods. Plan Theory 11(15):2070. https://doi.org/10.3390/plants11152070

Owomugisha G, Quinn J, Mwebaze E, Lwasa J (2014) Automated vision-based diagnosis of Banana bacterial wilt disease and black Sigatoka disease. In Booysen, MJ. (ed). 2014. Proceedings

of the 1rst international conference on the use of Mobile information and communication technology (ICT). in Africa UMICTA 2014
Ramírez H, Rey JC, Martinez G (2016) Relationship between the incidence of banana wilt and properties of a lacustrine soil. ACORBAT Meeting, 2016. Miami USA. Memories
Rey JC, Martínez G, Pizzo N, Micale E, Fernández N (2016) Susceptible areas to false Panama disease in Cavendish bananas. In: In the central region of Venezuela. In Resúmenes de Ponencias, pp 21, ACORBAT meeting, 2016, Miami USA
Rey JC, Martínez-Solórzano G, Ramírez H, Pargas-Pichardo R (2020) Relación de las condiciones agroecológicas de un lote de planicie lacustrina con la marchitez del banano Cavendish en Aragua, Venezuela. Agronomía tropical 70: 1–12. http://doi.org/https://doi.org/10.5281/zenodo.4346252
Sabadell S (2003) Etiología y epidemiologia del falso mal de Panamá en Canarias. Tenerife, España, Universidad Autónoma de Barcelona, Tesis Dr, p 286
Sangeetha T, Lavanya G, Jeyabharathi D, Rajesh Kumar T, Mythili K (2020) Detection of pest and disease in banana leaf using convolution random Forest. Test Eng Manag 83(2):3727–3735
Serrano E (2003) Relationship between functional root content and banana yield in Costa Rica. In Turner DW, Rosales FE (Eds.). Banana root system: towards a better understanding for its productive management. Presentado en el International symposium, San Jose, Costa Rica, 3-5 November 2003
Turner DW, Fortescue JA, Thomas DS (2007) Environmental physiology of the bananas (*Musa* spp.). Braz. J Plant Physiol 19(4):463–484
Vipinadas MJ, Thamizharasi A (2016) Banana leaf disease identification technique IJAERS 3(6):120–124
Ye HC, Huang WJ, Huang SY, Cui B, Dong YY, Guo AT, Ren Y, Jin Y (2020) Identification of banana fusarium wilt using supervised classification algorithms with UAV-based multi-spectral imagery. Int J Agric Biol 13(3):136–142. https://doi.org/10.25165/j.ijabe.20201303.5524
Yuan J, Wen T, Zhang H, Zhao M, Penton CR, Thomashow LS, Shen Q (2020) Predicting disease occurrence with high accuracy based on soil macroecological patterns of fusarium wilt. ISME J 14(12):2936–2950. https://doi.org/10.1038/s41396-020-0720-5

Chapter 5
Identification of the Main Soil Variables Correlated with Banana Productivity

Abstract The application of conventional agricultural techniques and high-cost inputs has not been sufficient to reverse the trend toward reduced banana yields in Venezuela. This has also been linked to the worsening of soil properties in banana plantations. This chapter aims to identify the main edaphic variables most correlated to banana productivity in Venezuela and explore the development of an empirical correlation model to predict this productivity based on soil characteristics. Six agricultural fields located in two of the main banana production areas of Venezuela were selected. Additionally, a productivity index (PI) based on three different biometric data on plant productivity was proposed (see Chap. 2). Thus, five multiple linear regression models were estimated, using the stepwise regression method. Subsequently, a performance analysis was used to compare the prediction quality range and the error associated with the number of soil variables selected for the proposed models. The selected model included the following soil variables: Mg, penetration resistance, total microbial respiration, bulk density, and omnivorous free-living nematodes. These variables explain the PI with an R^2 of 0.55, the mean absolute error (MAE) of 0.8, and the root mean squared error (RMSE) of 1.0. The regression model obtained allowed the development of an easy-to-implement methodology for the rest of the banana areas of Venezuela and Latin America, which when further validated, it would represent a tool for sustainable land management.

Keywords Banana · Free-living nematodes · Penetration resistance · Bulk density · Soil quality · Total microbial respiration

5.1 Introduction

Banana (*Musa* spp.) is an essential food source for a large part of the world's population. Its annual world production grew at a rate of 3.7% from 2000 to 2015, increasing from 68.2 million tons in 2000 to 117.9 million tons in 2015. Banana productivity levels vary depending on the producing country and banana variety. According to

B. O. Olivares Campos, *Banana Production in Venezuela*, The Latin American Studies Book Series, https://doi.org/10.1007/978-3-031-34475-6_5

FAO (2019) in the Venezuelan territory, there were 30,544 productive hectares of bananas (*Musa* AAA) in 2017, with a production of 424,649 tons destined for both the local market and for export, with an average yield of 13.91 tons ha^{-1}. The "Inter-American Network of Academies of Sciences" (IANAS 2017) recognizes that banana production in Venezuela is stagnant due to the combination of a shortage of agricultural inputs (fertilizers and agrochemicals), difficult access to foreign exchange to meet internal demand and external supplies of machinery or equipment, and also due to the absence of sustainable agricultural practices over time.

Conventional cultivation of bananas has been noted as driver of soil degradation, resulting in the loss of natural fertility due to inappropriate practices, demanding intensive use of agrochemicals aimed to maximize yield and enlarge cultivated areas (Viloria et al. 2003). In banana plantations in Venezuela, the most common practices are intensive tillage, monoculture, mechanization, and chemical fertilization characterized by high doses of fertilizers (Rey et al. 2006; Delgado et al. 2010a). Previous studies in banana and other tree crops have shown a direct relationship between the loss of soil quality and the reduction in crop productivity due to inappropriate conventional production system (e.g., Segura et al. 2015; Calero et al. 2018). Within this field, the interactions among soil properties, yield, and yield quality are still an active field of research with remaining open questions. For instance, Vega-Ávila et al. (2018) identified a set of variables (amylase, cellulase, and xylanase activities, bacterial abundance, microbial biomass and water content, pH, and electric conductivity) to be used in future studies relating soil quality and grape quality. Sometimes, these studies tried to identify the relationship between soil properties modified by specific agronomic practices and yield and plant physiological responses, as in the case of Martínez et al. (2018), which explored the effect of organic amendments in grapevines on soil quality and grapevine root development.

Despite these studies, more complete information on optimum microbiological, physical, and chemical conditions for banana roots and their relationship to banana plant development and productivity is lacking, as well as a deeper understanding of the relationship between soil properties and banana productivity in the main growing areas of Venezuela. So far, the methods used to measure the productive potential of banana soils in Venezuela have been based on general soil physical and chemical properties, geomorphology, topography, and climate (Delgado et al. 2010b; Rey et al. 2009) while rarely considering soil quality and its interactions with the rhizosphere (Rodríguez et al. 2006). This will contribute to facilitate the delimitation of zones with different soil qualities on which to apply differentiated management practices within the same banana plantation, a subject which is of great interest to farmers.

This manuscript presents a study aimed to validate the hypothesis that it is possible to delineate areas of different productivities within banana fields in the two of the main banana-producing areas in Venezuela (Aragua and Trujillo States) using an index derived from easily measured soil properties with a straightforward agronomic and environmental interpretation.

5.2 Results

5.2.1 *Productivity Index*

In the case of A1, T1, and T2, the results found in the aerial part of the crop (Table 5.1) show that the extreme levels (HH and LL) presented contrasting differences; however, the productivity levels characterized as HL and LH were similar. This is how the vigor of the plantation is greater in the lot designated as high productivity (HH) and progressively decreases until the lot of low productivity (LL). The populations of the different productivity lots were quite homogeneous, ranging between 1924 and 2039 plants/ha in A1.

Regarding the population of plants in T1, the tendency of the low-productivity lot (LL) to present higher populations than in the high-productivity lot (HH) is observed. For bunch weight, high-productivity lots were slightly higher than designated low-productivity lots. It should be noted that this bunch weight value is an estimate of a prediction equation based on the number of hands present in the bunch, which can, in many cases, overestimate the real value of the bunch weight, especially in lots of low productivity where the number of hands is like that of other levels; however, the filling fruit can be substantially different.

The results in relation to the aerial part of the crop in A2, A3 and A4 fields (Table 5.2) indicate that the greatest vigor of the plantation was markedly presented at the HH level in all the parameters evaluated. The differences between the HH and LL levels are considerable from the standpoint of crop vigor. The banana plant population was lower at high-productivity levels in A2 and A4 fields.

Figure 5.1 shows the results of the PCA for the first two principal components (PC1 and PC2) that account for most of the variance in biometric data (96.83%) on plant productivity. PC1 and PC2 explain 74.86% and 21.97% of the total data variability, respectively. In our study, only PC1 was selected to propose the PI, since this PC better explains the behavior of the variables: circumference pseudo-stem of the mother plant at 1 m in height and number of hands per bunch, as well as clusters of the different productivity classes (Fig. 5.1a). PC2 would be mainly explained by the height of the succession plant (Fig. 5.1a). Therefore, the PI will depend on the magnitude (weight and sign) of the vectors under study. It is also apparent in Fig. 5.1 how the different classes tend to overlap along PC2, especially for the larger spots, although there is a clear separation between the smaller and larger spots along this axis.

In general, the high-productivity sites are in the right side of Fig. 5.1b with higher scores on PC1, whose IP ranges from 0.06 to 2.70 (Fig. 5.1b). Sites located on the left panel, with lower PC1 values, would be the plots with the lowest productivity whose PI ranges from −4.21 to 0.11. The T2, A1, and A4 sites are those with the highest levels of banana productivity; these results contrast with the productivity observed in the A2 and A3 sites, which have the lowest productivity indices in the study (Fig. 5.1b).

Table 5.1 Average values of the main productive parameters in the banana sites (A1, T1, and T2): pseudo-stem circumference (cm) at 1 m height from the mother plant, height of the succession plant (cm) and number of hands per bunch (*n*), bunch weight (*n* = 20 plants), and plant population (plants/ha) in each level productivity

Fields	Productivity level[a]	Circumference (cm)	Height (cm)	N° of hands/bunch (*n*)	Bunch weight (kg)	Plant population (plants/ha)
A1	LL	58.74	159.87	7.08	29.46	2039
	LH	72.79	178.06	8.65	33.41	2005
	HL	73.86	189.66	9.02	34.36	1965
	HH	87.03	211.70	10.83	38.94	1924
T1	LL	62.50	168.16	9.09	34.54	1922
	LH	67.21	173.02	9.16	34.71	2015
	HL	65.95	201.07	8.78	33.76	1900
	HH	75.70	246.81	10.31	37.58	1852
T2	LL	64.41	163.62	8.11	32.05	1742
	LH	73.58	187.42	9.26	34.96	1885
	HL	76.11	199.95	9.37	35.24	1880
	HH	86.18	208.01	12.33	42.74	1520

[a]Classification nomenclature of banana soil productivity according to average fruit production: HH = high-high yield (≥40 t/ha/year); HL = high-low yield (35–40 t/ha/year); LL = low-low yield (≤30 t/ha/year); LH = low-high yield (30–35 t/ha/year)

Table 5.2 Average values of the main productive parameters in the banana sites (A2, A3, and A4): pseudo-stem circumference (cm) at 1 m height from the mother plant, height of the succession plant (cm) and number of hands per bunch (*n*) ($n = 20$ plants) and plant population (plants/ha) in each level productivity

Fields	Productivity level[a]	Circumference (cm)	Height (cm)	N° of hands/bunch (*n*)	Plant population (plants/ha)
A2	LL	51.25	70.11	6.37	2292
	HH	78.57	118.15	10.92	2166
A3	LL	60.15	62.62	7.65	1695
	HH	78.05	108.075	11.42	2330
A4	LL	70.41	134.75	9.58	2305
	HH	88.50	181.41	12.28	1872

[a]Classification nomenclature of banana soil productivity according to average fruit production: HH = high-high yield (≥40 t/ha/year); LL = low-low yield (≤30 t/ha/year)

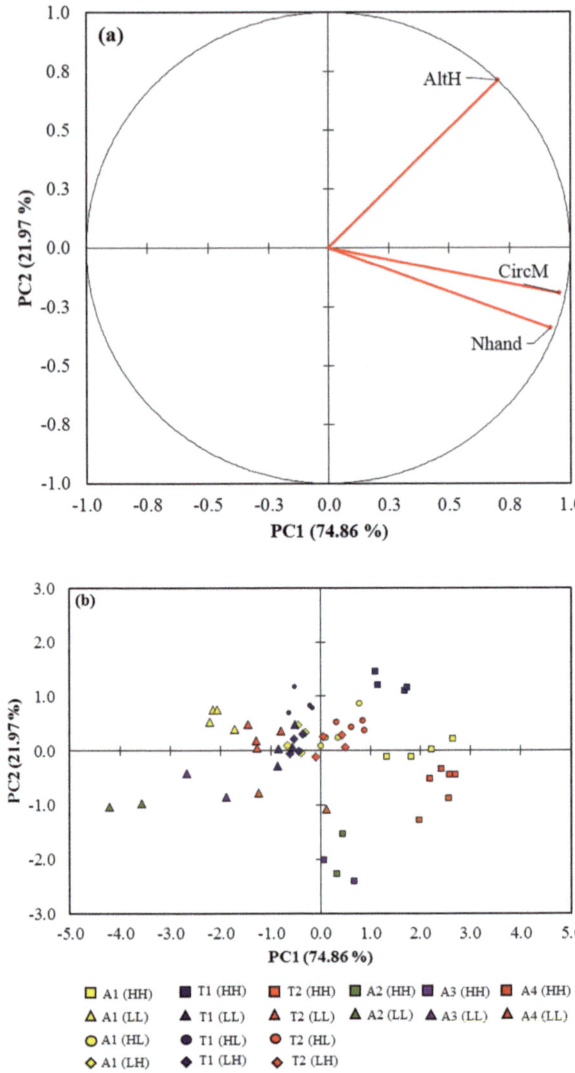

Fig. 5.1 (**a**) Projection of the circumference of the pseudo-stem of the mother plant (CircM), number of hands per bunch (Nhand), and the height of the succession plant (AltH) in principal components, indicating the relative weight of these variables into PC1 and PC2 through principal component analysis. Values in brackets in the axis legend indicate the proportion of variance explained by PC1 and PC2 through principal component analysis. (**b**) Bivariate representation of the results of the six banana sites (A1, A2, A3, A4, T1, and T2, with banana productivity according to the average fruit production of each plot evaluated (HH = high-high yield (≥40 t/ha/year); HL = high-low yield (35–40 t/ha/year); LL = low-low yield (≤30 t/ha/year); LH = low-high yield (30–35 t/ha/year)) within PC1 and PC2 (Olivares et al. 2020)

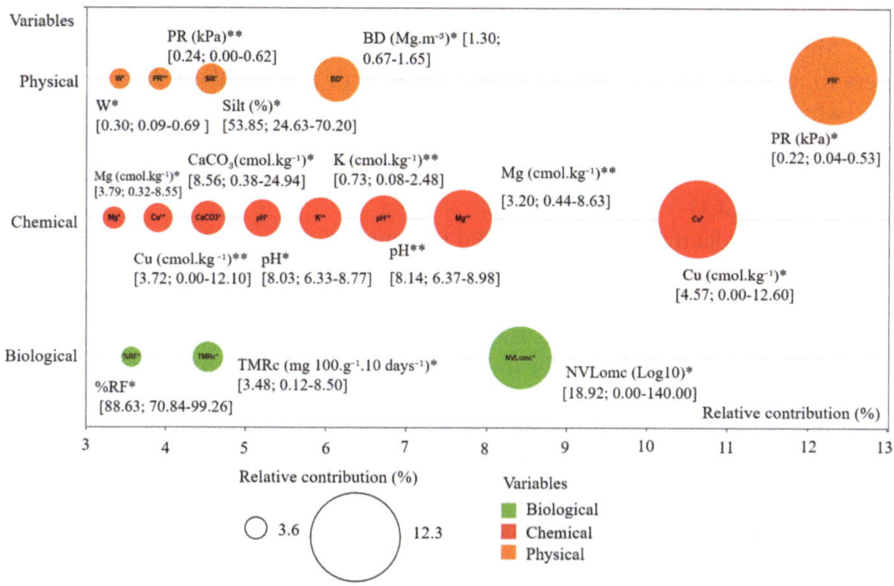

Fig. 5.2 Beeswarm plot of the relative contributions (%) [mean; lower and upper limit] of the main predictor variables using random forest with a tree complexity of 5 and a learning rate of 0.005. Note: omnivorous free-living nematodes (NVLomc), soil pH, magnesium (Mg), penetration resistance (PR), copper (Cu), total microbial respiration (TMRc), soil bulk density (BD), potassium (K), moisture index (W), and percentage of functional roots (% RF). *Data from soil horizon 1; **Data from soil horizon 2

5.2.2 Relative Importance of Predictor Variables

In total, 16 soil variables that accounted for the highest proportion of the variance of PI were selected in the random forest analysis (Fig. 5.2) including penetration resistance (PR), copper (Cu), magnesium (Mg), and soil pH from both soil horizons, while the variables omnivorous free-living nematodes (NVLomc), calcium carbonate ($CaCO_3$), silt, total microbial respiration (TMRc), moisture index (W), soil bulk density (BD), percentage of functional roots (% RF) were selected from data on soil horizon 1 only and potassium (K) from soil horizon 2. Figure 5.2 shows the distribution of soil variables in relative contributions (%), as a continuous dimension. Each soil property is represented by a point placed on the horizontal axis. The vertical dimension represented by the type of physical, chemical, and biological variable is used to avoid overlaps between circles, showing their distribution. The area of the variables shows in turn the size of the relative contributions (%).

The high-productivity soils in A1 field presented loamy textures, with a predominance of particles with an equivalent diameter between 2 and 50 µm, whose bulk density values were low (0.45–0.89 $Mg.m^{-3}$), which is to be expected, since they are soils of lacustrine origin, whose parent material provides such characteristic. For this reason, the total porosity values were quite high (63–78%). These characteristics allow these soils to have a moderate to moderately high infiltration rate. Similarly, the permeability measured by hydraulic conductivity was moderately to moderately high. Regarding the modulus of rupture, the values were very low, which indicates that they have a low dry cohesion. From the physical point of view, these soils did not present major limitations. The limitations for root development in these soils could be associated with chemical conditions, such as the presence of high $CaCO_3$ contents.

The low-productivity soils in A1 field presented medium textures on the surface to heavy in the subsurface layers, with a high predominance of particles with an equivalent diameter between 2 and 50 μm and higher content of particles with equivalent diameter <2 μm in the subsurface layers. Also, they presented high values of bulk density for this type of soil (0.94–1.28 $Mg.m^{-3}$), with low values of total porosity (53–65%). Infiltration and hydraulic conductivity values were low to moderate, which can limit the penetration and movement of water in the soil. In the same way, relatively high values of modulus of rupture were obtained. In these soils, the highest values of resistance to penetration were obtained, which could affect the normal development of banana roots.

Regarding the physical variables in T1, in general, strong limitations were observed, which are accentuated in the low-productivity lots. These limitations are fundamentally associated with susceptibility to compaction and surface sealing, as well as restrictions for the penetration and movement of water in the soil. In the same way, high levels of modulus of rupture were observed in the limiting range for the crop. The chemical variables most related to productivity levels were pH and low Ca/K and Mg/K ratios. The pH was very high at all productivity levels, which was related to high levels of Na and Fe deficiency. On the other hand, low Ca/K and Mg/K ratios cause nutritional imbalance and difficulty in the absorption of these items.

The high-productivity soils in T1 presented a higher number of bacterial and fungal populations, indicating a higher biological activity, an aspect that is related to the higher content of microbial carbon in these soils. However, total respiration was lower in high-productivity soils than in low-productivity soils. Biological variables presented a less consistent relationship with banana productivity levels than in the case of physical and chemical variables. However, a higher population of bacteria and fungi stands out in the good soils of most of the sites, associated with a higher presence of microbial carbon, as well as the highest proportion of phytonematodes in low-productivity soils.

The physical properties of the soil in T2 presented greater limitations in the level of low productivity in terms of the penetration and movement of water in the soil. Likewise, high values of resistance to penetration and modulus of rupture were presented, even when the bulk density is not limiting according to the textural classification of these soils. These characteristics are associated with problems in the development and distribution of the root system and consequently on crop yield. The chemical variables most related to productivity levels were pH and low Ca/Mg, Ca/K, and Mg/K ratios. The pH was very high in poor soils, which was related to Fe and Zn deficiency. On the other hand, low Ca/Mg, Ca/K, and Mg/K ratios cause nutritional imbalance and difficulty in the absorption of these elements.

5.2.3 Multiple Linear Regression

A functional relationship was established between the PI and the statistically significant variables ($P < 0.05$). A total of nine variables among those identified in the random forest analysis were included in the model that showed the highest goodness of fit ($R^2 = 0.69$), indicating that there is a good degree of association between the PI and the predictor variables in the model (Eq. 5.1). The variables in Eq. 5.1 are described in Fig. 5.3. For this model, the coefficients are robust, being within the bootstrap confidence interval, so the stability of the model is assured, and the results can be generalized.

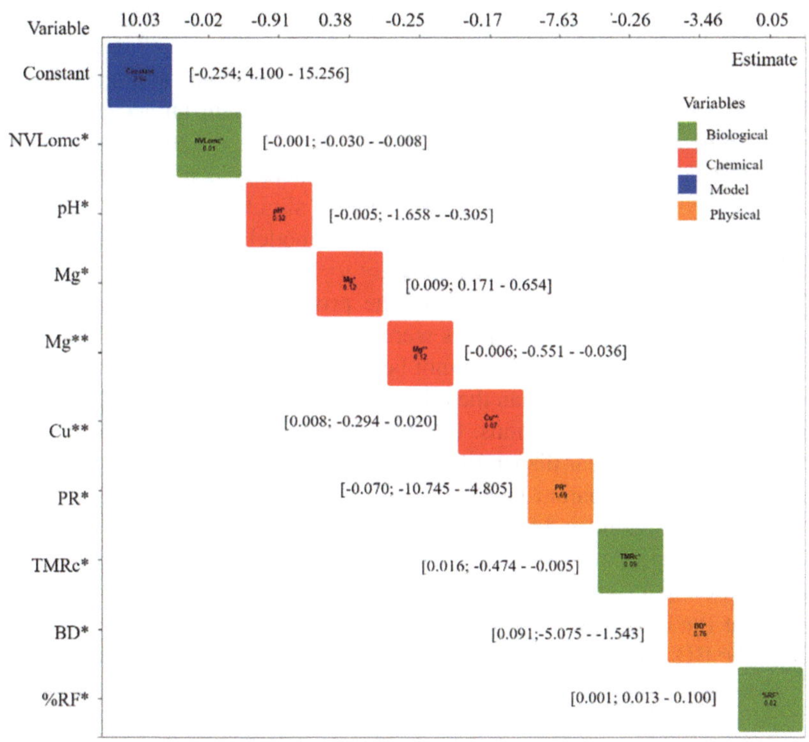

Fig. 5.3 Bootstrap coefficients (estimate in *y* axis, standard error inside the box) [bias and 95% percentile confidence intervals] for bootstrap of the model parameters of the population mean based on the 2000 bootstrap samples. Omnivorous free-living nematodes (NVLomc); soil pH, magnesium (Mg), penetration resistance (PR), copper (Cu), total microbial respiration (TMRc), soil bulk density (BD), and percentage of functional roots (% RF). *Data from soil horizon 1; **Data from soil horizon 2

$$\begin{aligned}\mathrm{PI} = 10.03?0.02\left(\mathrm{NVLomc}^{?}\right)?0.91\left(pH^{?}\right)+0.38\left(\mathrm{Mg}^{?}\right)?0.25\left(\mathrm{Mg}^{??}\right) \\ +0.17\left(Cu^{??}\right)?7.63\left(PR^{?}\right)?0.26\left(\mathrm{RMTc}^{?}\right)?3.46\left(BD^{?}\right)+0.05\left(\%RF^{?}\right)\end{aligned} \tag{5.1}$$

5.2.4 *Performance Analysis of the Selected Soil Variables*

Performance analysis was carried out for five models developed from the model described in Fig. 5.3 reducing the number of variables. For the first model, nine variables were used according to Eq. 5.1. From the more complex model (model 1 in Fig. 5.4), at each iteration, the variable whose level of significance was lower was gradually discarded from the model. Figure 5.4 shows the differences among regression models in terms of R^2, MAE, and RMSE. The results of this analysis showed that as the number of variables in the model decreases, R^2 decreases to the point where the model performed poorly (model 5, which only explains about 25% of the total variability). In general, PI models 1–4 present an acceptable performance. The best performed model explained 70% of the variation in the measured PI. Models 1–3 had the lowest values of MAE (generally <0.8) and RMSE (generally ≤1.0 PI) (Fig. 5.4). The fourth model showed slightly higher MAE and RMSE values than the previous models, while the fifth model had the highest MAE and RMSE values (1.05 and 1.30, respectively).

To visualize the predictive ability of the proposed models, the values of the observed and predicted PI of all plots were plotted (Fig. 5.5). In Fig. 5.4, model 1 presented the highest R^2 value. It cannot be stated that this is the best model, since although it has an R^2 greater than the other models, there may be an overadjustment or simply not being parsimonious. For model 3 (Fig. 5.5), the data was clustered near the 1:1 line, a similar dispersion was observed between the data observed and predicted. It is also apparent how model 5 presented a poor predictive capability with the lowest R^2 value and far from the 1:1 line.

Results suggested that the PI may have a strong correlation with BD given the large reduction in R^2 when this variable was not included in model 5. It is important to note that the error (MAE: 1.05 and RMSE: 1.30) associated with the model 5 (without the BD variable) was the highest, while the models that considered the BD had a more acceptable, and similar, performance (Fig. 5.4). According to all previous results, the selected model would be model 3 (Eq. 5.2), which is composed of five variables that explain the PI with an R^2 of 0.55, a MAE of 0.8, and a RMSE of 1.0.

$$\begin{aligned}\mathrm{PI} = 9.03+0.19\left(\mathrm{Mg}^{?}\right)?8.78\left(PR^{?}\right)?0.40\left(\mathrm{TMRc}^{?}\right)?4.70\left(BD^{?}\right) \\ +0.02\left(\mathrm{NVLomc}^{?}\right)\end{aligned} \tag{5.2}$$

where Mg = magnesium, PR = penetration resistance, TMRc = total microbial respiration, BD = soil bulk density, and NVLomc = omnivorous free-living nematodes. *All soil variables were obtained from soil horizon 1.

Fig. 5.4 Circular dendrogram of performance analysis represented by the coefficient of determination of the five models hierarchically structured with a radial tree structure, where the root node is in the center with the hierarchies moving outward divided into branches: physical (P), chemical (C), and biological (B) properties. The area of nodes was used to encode the R^2 as a quantitative dimension. Note: magnesium (Mg), copper (Cu), percentage of functional roots (% RF), soil pH, soil bulk density (BD), penetration resistance (PR), total microbial respiration (TMRc), and omnivorous free-living nematodes (NVLomc). The root mean squared error (RMSE), the mean absolute error (MAE), and the coefficient of determination (R2). *Data from soil horizon 1; **Data from soil horizon 2

The purpose of this analysis was to look for parsimony using fewer variables (losing some precision) but that would best represent the PI. Although the R^2 of 0.55 is not completely satisfactory, it is an effective indicator at the field level, where you can usually see coefficients of variation close to 40% (regardless of which variable is being measured). Using five variables instead of nine (model 1) will allow the banana producers to simplify the number of soil analysis to perform, with the consequent saving of money and time, and the making more feasible for field implementation.

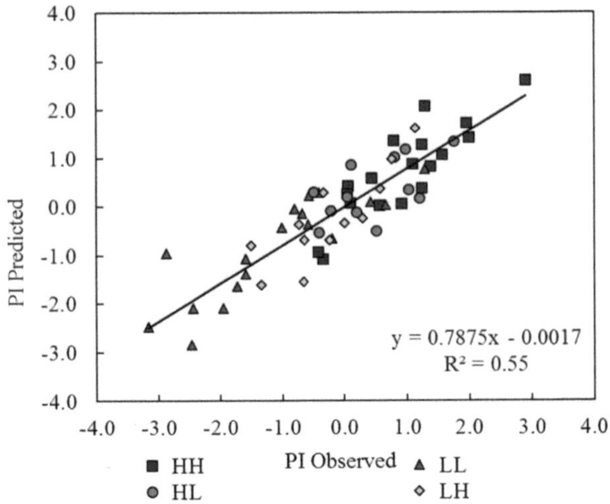

Fig. 5.5 Results of productivity index (PI) observed in the six banana sites with banana productivity according to the average productivity index of each plot evaluated (HH = high-high yield (≥40 t/ha/year); HL = high-low yield (35–40 t/ha/year); LL = low-low yield (≤30 t/ha/year); LH = low-high yield (30–35 t/ha/year)) versus the predicted PI of the best selected model in the study (model 3). The black line indicates the correlation between that observed and that predicted (Olivares et al. 2020)

5.2.5 The MLR Solution by Bootstrapping

Figure 5.6 shows the bootstrapping performed on model 3, which indicates that the coefficients have a small bias (i.e., very little difference between the mean value of the coefficients in the 2000 resamples and the original coefficients). The bootstrap coefficients are still significant, and 95% of the values of the resampled coefficients are within the range (lower-upper) of the estimated confidence intervals. This shows that the model is stable and might be generalizable, so the results can be generalized for similar situations. These results support the decision to select model 3 instead of model 1 composed of nine variables, precisely due to the parsimony criterion.

5.3 Discussion

5.3.1 The Role of Soil Quality in Productivity

Nowadays, there is a great interest from banana farmers in Venezuela, among other producing countries, to optimize input use in plantations. For this, characterizing the productive efficiency of banana plots will facilitate the delimitation of zones for differentiated management (Castañeda et al. 2010). This study identified the soil

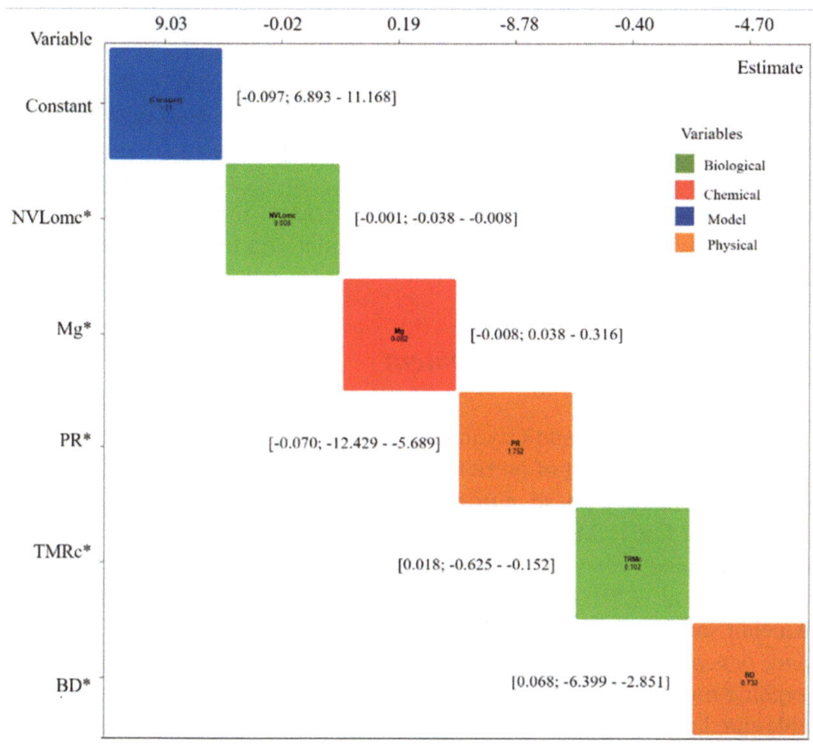

Fig. 5.6 Bootstrap coefficients (*B*) in *y* axis and standard error inside the box [bias and confidence interval (95%)] for regression coefficients of model 3 for the population mean based on the 2000 bootstrap samples. Omnivorous free-living nematodes (NVLomc), magnesium (Mg), penetration resistance (PR), total microbial respiration (TMRc), and soil bulk density (BD). Note: All soil variables belong to horizon 1

properties most associated with banana productivity in two producing areas if Venezuela to explore the potential development of empirical simple models that could help in predicting the banana productivity by estimating a PI as a function of soil characteristics. In these conditions, the three biometric variables associated with the PI used (circumference of the pseudo-stem of the mother plant, number of hands per bunch, and the height of the succession plant) confirmed the prior designation of each sampled area in different categories according to the production of each of those spots, indicating its validity as proxies of yield. This is in line with previous studies that also have established good accuracy in the estimation of banana productivity using biometric characteristics of the crop (Sumner 1990; Rodríguez et al. 2004, 2005; Segura et al. 2015) but not in the regions covered by our study.

We have demonstrated that it is possible to develop different models to characterize the PI in banana plantations based on soil characteristics and provided guidance for selecting the best choice from a stakeholder's perspective. In this approach, model 1, although showing better correlation with PI, was ruled out as a practical alternative for estimating banana productivity because it included too many soil variables—nine—and consequently would be more difficult to be implemented at the field level. Therefore, the best models to be implemented at the field level in terms of information needed and ability of PI prediction were models 2 and 3 (with six and five soil variables included in each model, respectively). Model 3 was finally selected due to similar performance to model 2 and because it was based on only five soil variables: magnesium, penetration resistance, total microbial respiration, soil bulk density, and omnivorous free-living nematodes from soil horizon 1.

5.3.2 Soil Chemical Management

The findings of Martínez and Cayón (2011) indicate that in the vegetative phase of bananas, the accumulation of nutrients such as N, P, Ca, B, K, and Mg is required, while in the productive phase, nutrients such as K and N are needed. The progressive increase of Ca, P, Mg, and K (g/plant) during the floral differentiation phase could be explained because Ca is essential in the formation of cell walls.

Delgado et al. (2010b) also estimated an index of quality and health for Venezuelan soils, in Aragua and Trujillo, associated with pH, magnesium, copper, % sand, penetration resistance, free-living nematodes, total microbial respiration, and organic matter. Although they did not link it to a productivity index. These soil variables are largely consistent with the results reported in our study. It is worth noting that variables that were significant in our model to estimate the PI are not usually considered by Venezuelan producers, especially for banana production, and so they provide additional insight for improving agronomical practices.

For instance, magnesium is an element almost not considered in fertilization programs, low attention is played to physical conditions (soil bulk density and penetration resistance), and negligible consideration is given to biological aspects, such as free-living nematodes, which have shown to be an adequate variable to estimate productivity. There is no need to emphasize that the kind of models linking soil quality and yield are site specific, as indicated previously by Delgado et al. (2010b), Villarreal-Núñez et al. (2013), and Rey et al. (2009). Nevertheless, these models aimed to identify variables related to banana productivity use variables similar to the ones identified in our study: pH, magnesium, copper, sand content, penetration resistance, total microbial respiration, organic matter, radical weight, total bacteriophages' free-living nematodes, and abundance of *Trichoderma.*

As noted above, our analysis can provide an overall insight into key variables that might be hampering banana productivity. So, Mg levels were moderate to low in all the plots sampled in our study but always lower in the high-productivity plots (Table 5.3, Appendix), indicating that there may be problems of nutritional balance and absorption of these nutrients by part of the plants. Our results have shown a clear correlation of Mg with the PI in the study areas, even though the literature indicates that Mg requirements for crop growth are smaller than Ca requirements. Fageria (2009) suggests that, in these areas, there is a response to differences in

Table 5.3 Mean values ± standard deviations for the variables selected in model 3: magnesium (Mg) content, penetration resistance (PR), total microbial respiration (TMR), soil bulk density (BD), and omnivorous free-living nematodes (NVLomc) in the six banana sites included in the study

Fields	Productivity level[a]	Mg ($cmol.kg^{-1}$)	PR (kPa)	TMRc (mg $100.g^{-1}$)	BD ($Mg.m^{-3}$)	NVLomc (Log10)
Aragua state						
A1	HH	7.22 ± 0.90	0.35 ± 0.05	5.78 ± 1.67	0.79 ± 0.06	2.75 ± 1.25
	HL	5.44 ± 1.37	0.32 ± 0.03	7.61 ± 0.75	0.77 ± 0.09	9.25 ± 8.53
	LL	7.23 ± 0.90	0.41 ± 0.08	5.34 ± 0.72	1.26 ± 0.03	28.00 ± 13.85
	LH	6.82 ± 0.61	0.29 ± 0.07	6.28 ± 1.51	1.14 ± 0.05	25.75 ± 16.85
A2	HH	1.56 ± 0.11	0.11 ± 0.02	0.29 ± 0.26	1.45 ± 0.02	130.00 ± 14.14
	LL	1.76 ± 0.01	0.30 ± 0.01	2.67 ± 0.02	1.40 ± 0.10	35.00 ± 7.07
A3	HH	0.97 ± 0.38	0.13 ± 0.01	1.61 ± 0.57	1.54 ± 0.05	5.00 ± 7.08
	LL	1.00 ± 0.32	0.17 ± 0.00	2.53 ± 0.67	1.36 ± 0.11	10.00 ± 14.14
A4	HH	0.96 ± 0.09	0.07 ± 0.04	2.83 ± 0.01	1.35 ± 0.04	40.00 ± 14.13
	LL	1.83 ± 0.03	0.23 ± 0.01	4.30 ± 0.07	1.29 ± 0.01	70.00 ± 70.70
Trujillo state						
T1	HH	5.15 ± 0.28	0.23 ± 0.12	1.34 ± 0.62	1.42 ± 0.07	2.00 ± 2.31
	HL	4.92 ± 0.30	0.23 ± 0.03	1.44 ± 1.30	1.50 ± 0.06	4.00 ± 3.27
	LL	4.50 ± 0.71	0.16 ± 0.01	2.26 ± 1.28	1.53 ± 0.08	12.00 ± 5.66
	LH	5.85 ± 1.00	0.24 ± 0.08	3.11 ± 0.74	1.54 ± 0.07	10.00 ± 10.07
T2	HH	1.49 ± 0.76	0.09 ± 0.04	1.85 ± 0.81	1.26 ± 0.10	12.00 ± 0.01
	HL	0.66 ± 0.37	0.13 ± 0.05	2.38 ± 1.34	1.38 ± 0.12	9.00 ± 1.10
	LL	1.82 ± 0.32	0.22 ± 0.02	3.56 ± 1.50	1.33 ± 0.03	12.00 ± 9.70
	LH	1.69 ± 0.52	0.12 ± 0.02	4.12 ± 2.19	1.29 ± 0.14	12.00 ± 9.79

All soil properties correspond to horizon 1

[a]Classification nomenclature of banana soil productivity according to average fruit production: HH = high-high yield (≥40 t/ha/year); HL = high-low yield (35–40 t/ha/year); LL = low-low yield (≤30 t/ha/year); LH = low-high yield (30–35 t/ha/year) (Olivares et al. 2020)

available Mg, as noted in other banana-growing regions because it is a neglected element in fertilization. An example of previous findings in this direction is that of Irizarry et al. (1990) in Ultisol soils in Puerto Rico that indicated a significant increase in yield of bananas to Mg fertilization.

5.3.3 Soil Physical Management

It is important that a balance is generated between the components of the soil (air, water, minerals, and organic matter) so that the crop can produce larger bunches, with an expectation of higher productivity. According to Lahav and Turner (1989), 70% of the bunch weight will be determined by the physical properties of the soil. The physical preparation of the soil involves subsoiling work, with machinery for those areas that are compacted, which favors macroporosity, water absorption by the plant, and root development (Rodríguez and Lobo 2004). In some areas with compacted soils, oxygen is lacking, nutrient absorption is deficient, and the balance between oxidation and reduction is not achieved (Pulido-Moncada et al. 2009). The plant will not be able to take the carbon it requires for the photosynthesis process; therefore, it will not have good nutrition, and it will have a smaller size, causing stagnation and dehydration. Therefore, a compacted soil results in small and poor-quality bunches (Olivares et al. 2022).

In our study, the physical variables most related to crop productivity levels were BD and PR. According to Blanco-Sepúlveda (2009), there is a direct relationship between the PR and the BD. In the light of the results obtained in this work, the soil BD can be used as a relevant soil property to estimate the potential banana productivity in different producing areas in Venezuela. The high BD values observed in some plots are associated with the lacustrine origin of these, which can affect the normal development of banana roots. Bananas show a high sensitivity to root damage and decreased yields due to soil compaction (Villarreal-Núñez et al. 2013), which is also confirmed by studies of Rey et al. (2006) in Venezuela and of Pattison et al. (2005) in Queensland (Australia), noting that most of the problems related to soil quality and low banana plantation productivity were related to some of the physical properties of the soil restricting root development. Thus, the weight of the roots and shoots of the banana decreases significantly as the BD of the soil increases. Similarly, banana at high soil compaction presents a significant decrease in the total number of leaves, the average leaf area, the total leaf area, the diameter of the pseudo-stem, and the height of the succession plant (Villarreal-Núñez et al. 2013). Drastic changes in soil aggregation, in the presence of expandable clays or soil compaction associated with less porosity and poor drainage, could damage roots and affect banana production (Araya and Blanco 2001; Castañeda et al. 2010). Textural extremes (clay and sand) are also factors that limit the growth of the banana root (Aguirre et al. 2012). Indeed, for heavy textures, where there is high bulk density and resistance to root penetration, water availability and nutrient absorption by crops are highly restricted.

Compaction is one of the soil degradation processes that reduces productivity mainly in herbaceous plants such as banana (*Musa acuminata* spp.). The study by Bustamante León et al. (2018) in banana soils from Ecuador demonstrated the relationship of BD with the levels of OM and soil sand, whose values indicated a higher degree of compaction. Also, the higher content of OM and fine particles increase the capacity of soils to store water, causing the soil to have a lower probability of compacting.

5.3.4 Soil Biological Management

The inclusion of biological soil parameters such as the abundance of free-living nematodes and microbial abundance and activity in soil quality studies, as we have done in this study, can provide significant knowledge of the link between the ecological role of soil fauna (Vega-Ávila et al. 2018) and effect of these organisms on banana productivity. The large number of culturable microorganisms and nematodes found in the soil associated with the banana rhizosphere in soils of the two localities sampled in our study could be attributed to the changes made by the plant roots in the surrounding rhizosphere microenvironment, resulting in higher populations in the rhizosphere of the most producing trees, something also found by Segura et al. (2015) in banana plantations in Costa Rica. Soil nematodes are a group of invertebrates of high ecological importance, which have attributes that make them valuable tools as biological indicators (Salas and Achinelly 2020). In the sampled soils in our study, free-living fungivore nematodes were predominant, followed by omnivores and lower proportions of bacteriophage and predatory microorganisms (*data not shown*). The low proportion of this last trophic group could reflect the presence of soil disturbance due to intensification and poor management of soils in the experimental spots evaluated. Our results agree with those of Castilla-Díaz et al. (2017), which found the genus c.f. Dorylaimida (omnivorous) as the most abundant nematode in Colombia. Millán et al. (2016) showed that greater diversity and richness of nematode populations, mainly phytophagous and omnivorous, are observed in medium-textured soils, whereas soils of heavier texture presented more abundance of the genus c.f. Dorylaimida in areas in which banana is cultivated. Other studies have found that omnivore and predator nematodes are less abundant in disturbed soils or intensive banana plantations (Ferris et al. 1996; Bautista et al. 2015; Olivares et al. 2022).

The metabolic activity of soil microorganisms is responsible for important processes such as mineralization and humification of soil organic matter (Martínez et al. 2018; Trujillo-Narcía et al. 2019). In our research, high-productivity plots (HH) showed low microbial respiration in all the sites evaluated, and in fact, soil organic carbon was not among the soil properties selected to be part of the predictive model of banana productivity based on soil properties. These results are consistent with the experiences of Gauggel et al. (2005) and González-Pedraza et al. (2014) that indicated the lower amount of microbial carbon and total organic carbon observed in Venezuelan soils with high-productivity plants. This indicates that specific strategies to increase soil organic carbon content in banana fields for the provision of other ecosystem services, such as, for instance, carbon sequestration, need to be identified since farmers might find little incentive to increase soil organic carbon if their link to yield is not straightforward in the short or medium term. The lower concentration of microorganisms observed in the roots of high-productivity plots (HH) could be a result of biochemical relationships between different nutrients (Al, Cu, Zn, Hg, and Ag) and the other microorganisms; these elements can be part of important enzymatic processes in the rhizosphere and root tissues. In typical Latin American areas of Musaceae production, such as Venezuela, Colombia, and Costa Rica, several authors (Delgado et al. 2008, 2010c; Hernández et al. 2007) have reported similar results to those presented in this research. The differences found in the values of total soil microbial respiration between the plots of high and low productivity could be due to differences in the composition of the microbial communities in the rhizosphere area in both soils, which deserves further study.

A study by Durango Cabanilla et al. (2017) stated that the contribution of organic amendments increased the microbial activity of the soil, which was reflected in greater CO_2 produced, as a result of the metabolism of microorganisms. In addition, as time progresses, the effect of the amendments decreases, this being an indication to mention that continuous applications of organic amendments are required to stimulate microbial activities. He was also able to conclude that the organic amendments and the biostimulant significantly increased the agronomic variables of development (plant height, stem diameter) and therefore there is a greater amount of dry matter in the banana plant.

5.3.5 Integrated Soil Management

An analysis of foliar tissue is essential to be able to propose specific and precise technological alternatives, in the evaluated lots of the largest farms. In this way, the possible fertilization doses could be established more appropriately, not only based on the results of the physicochemical analysis of the soil but also based on the absorption and nutrition of the crop.

In this last aspect related to the population of microorganisms in various banana lots studied, the small population present in low-productivity lots has tried to maintain a certain environmental balance, as happens in other adverse environments. The quality and health of the soil are directly related to microbial activity and the efficiency that it has on the decomposition of soil organic matter and the cycle of nutrients, so the use of organic amendments of varied composition and objectively treated, in such a way as to favor the mineralization processes of the organic matter. This practice must be accompanied by others that favor sustainable land management.

Additionally, the use of biological inputs or beneficial microorganisms is suggested, adaptable to the conditions of the banana farms studied so that they stimulate the metabolic and enzymatic processes necessary in the agroecosystem. From the results obtained, it can be inferred that the microflora present in the banana rhizosphere does not meet the expectations of improving or sustainably maintaining the productivity and vigor of the banana plantation in Venezuelan localities in the long term.

The results in our research demonstrate the complex interactions between the chemical, physical, and microbiological variables of the soil and how they affect banana productivity. This information, and particularly the methodological approach to develop the model linking selected soil properties with banana productivity, is innovative and new in the banana cultivation of Venezuela. It shows that the microbial respiration and the concentration of free-living nematodes of the banana soil are suppressed in areas with high productivity (i.e., HH plots). Practices to increase and maintain soil quality and stimulate microbiological activity in those soils could have a positive effect on agricultural banana production, not only for low-productivity fields but also for sustainable use of high-productivity lots.

5.4 Conclusions

The findings of this chapter indicate that the model has the ability to predict and establish a direct relationship between banana productivity based on the biometric data of the plant and the edaphic component, resulting in five soil properties usually represented as soil quality indicators (Mg, resistance to penetration, total microbial respiration, bulk density of the soil, and free-living omnivorous nematodes), which are important from the agronomic and environmental point of view.

Basically, the model reported in this chapter could be implemented in the field to reliably identify areas of high and low banana productivity, not only in banana-producing areas of Aragua and Trujillo, but can be extrapolated to areas whose agroecological conditions allow it, for example, areas of the states of Barinas, Monagas, Sucre, Yaracuy, and Zulia, logically with a correct validation in the field. Also, the identification of these physical, chemical, and biological properties of the soil can provide guidance to commercial farms and small and medium producers on the selection of sustainable soil management practices, without neglecting the natural variability of soils.

The growing need to obtain a higher banana production per hectare requires comprehensive soil management practices, which must consider not only the application of fertilizers but also the development of soil loosening works and initiatives that favor root development. A healthy soil with good quality indicators favors the production of larger bunches and strengthens the plant's defense against foliar diseases. However, the results obtained in this chapter are possibly the result of conventional and intensive management of the agroecosystem, characterized by a monoculture and by edaphoclimatic conditions that are not favorable for the soil microflora in low-productivity banana areas.

Appendix

References

Aguirre ES, Piraneque N, Menjivar Flores CJ (2012) Relationship between soil and edaphoclimatic properties and the Black Sigatoka incidence (*Mycosphaerella fijiensis* Morelet) in the Banana Region of Magdalena – Colombia. Rev Investig Agrar Ambient 3:1–13. https://doi.org/10.22490/21456453.970

Araya M, Blanco F (2001) Changes in the stratification and spatial distribution of the banana (Musa AAA cv. Grand Naine) root system of poor regular and good developed plants. J Plant Nutr 24:405–405. https://doi.org/10.1081/PLN-100107306

Bautista LG, Bolaños MM, Massae Asakawa N, Villegas EB (2015) Plantain *Musa* AAB simmonds phytonematodes response to soil integrated management strategies and nutrition. Rev Luna Azul 40:69–84. https://doi.org/10.17151/luaz.2015.40.6

Blanco-Sepúlveda R (2009) Relationship between bulk density and mechanical resistance as an indicator of soil compaction. Agrociencia 43:231–239

Bustamante León M, Chabla Carrillo J, Barrezueta Unda S (2018) La densidad y humedad crítica como indicadores de la compactación de suelos cultivados con banano. Agroecosistemas 6(1):168–174

Calero J, Aranda V, Montejo-Raez A, Martin-Garcia JM (2018) A new soil quality index based on morpho-pedological indicators as a site-specific web service applied to olive groves in the province of Jaen (South Spain). Comput Electron Agric 146:66–76. https://doi.org/10.1016/j.compag.2018.01.016

Castañeda DA, Jaramillo DF, Cotes Torres JM (2010) Components of the spatial variability in site-specific management in banana. Pesqui Agropecu Bras 45:836–845. https://doi.org/10.1590/S0100-204X2010000800009

Castilla-Díaz E, Millán-Romero E, Mercado-Ordoñez J, Millán-Páramo C (2017) Relation of soil parameters on the diversity and spatial distribution of free-living nematodes. Tecnol Marcha 30:24–34. https://doi.org/10.18845/tm.v30i3.3270

Delgado E, Gómez N, González O, Marín C (2008) On farm evaluation of effect of high plant density on plantain (*Musa* AAB cv. Subgroup Horn plantain), Obispo County, Barinas state, Venezuela. Rev Fac Agron (LUZ) 25:603–616

Delgado E, Rosales F, Trejos J, Villalobos M, Pocasangre L (2010a) An index of soil quality and health for banana plantations in four countries of Latin America and the Caribbean. Bioagro 22:53–60

Delgado E, Trejos J, Villalobos M, Martínez G, Lobo D, Carlos Rey J, Rodríguez G, Rosales FE, Pocasangre LE (2010b) Determination of a soil quality and health index for banana plantations in Venezuela. Interciencia 35:927–933

Delgado EJ, Carlos Rey J, Martínez G, Lobo Luján D, Sapuky M, Chacin M, Nuñez E, Pocasangre L, Rosales F (2010c) Propiedades del suelo que determinan el vigor del banano (Musa AAA) de producción local en Venezuela. XIX Reunión Internacional ACORBAT Medellín, Colombia, nov 8–12 (In Spanish)

Durango Cabanilla WD, Mite Vivar F, Carrillo Zenteno M, Cargua Chávez J, Lahuathe Mendoza B, Rivadeneira Moreira B, Moreira Zambrano V (2017) Evaluación de enmiendas orgánicas sobre la respiración microbiana del suelo y variables agronómicas en banano. J Sci Res 2(8):28–32

Fageria NK (2009) The use of nutrients in crop plants. CRC Press, Florida

FAO (2019) Banana market review and banana statistics 2018. Rome. http://www.fao.org/economic/est/est-commodities/bananas/en/. Accessed 12 Jan 2020

Ferris H, Venette RC, Lau SS (1996) Dynamics of nematode communities in tomatoes grown in conventional and organic farming systems and their impact on soil fertility. Appl Soil Ecol 3:161–175. https://doi.org/10.1016/0929-1393(95)00071-2

Gauggel CA, Sierra F, Arévalo G (2005) The problem of Banana root deterioration and its impact on production. In: DW, Rosales FE (eds) Banana root system: towards a better understanding for its productive management: proceedings of an international symposium. International network for the improvement of Banana and plantain, Montpellier, France, pp 13–22

González-Pedraza AF, Atencio J, Cubillan K, Almendrales R, Ramirez L, Barrios O (2014) Microbial activity in soils cultivated with plantain (*Musa* AAB plantain subgroup cv. Horn) with different vigor of plants. Rev Fac Agron (LUZ) 31:526–538

Hernández Y, Marín M, García J (2007) Response to the plantain (*Musa* AAB cv. Horn) yield as a function of the mineral nutrients and its phenological cycle. Part I. Growth and production. Rev Fac Agron (LUZ) 24:607–626

IANAS (Interamerican Network of Academies of Sciences. México) (2017) Retos y oportunidades de la seguridad alimentaria y nutricional en las Américas: El punto de vista de las Academias de Ciencias. IANAS, México (In Spanish)

Irizarry H, Rivera E, Rodriguez JA (1990) Response of bananas (*Musa-acuminata.* AAA) to magnesium fertilization in an Ultisol. J Agr U Puerto Rico 74:419–426

Lahav E, Turner DW (1989) Banana nutrition. IPI-Bulletin 7:1–33

Martínez AAM, Cayón SDG (2011) Dinámica del crecimiento y desarrollo del banano (*Musa* AAA Simmonds cvs. Gran Enano y Valery). Rev Fac Nal Agr Medellín 64(2):6055–6064

Martínez M, Ortega R, Janssens M, Fincheira P (2018) Use of organic amendments in table grape: effect on plant root system and soil quality indicators. J Soil Sci Plant Nutr 18:100–112. https://doi.org/10.4067/S0718-95162018005000501

Millán E, Castilla D, Millán C (2016) Comunidades de nematodos de vida libre del suelo y su correspondencia con la calidad. Ingenieria Reg 16:25–34. https://doi.org/10.25054/22161325.1282. (In Spanish)

Olivares BO, Araya-Alman M, Acevedo-Opazo C, Rey JC, Cañete-Salinas P, Kurina FG, Balzarini M, Lobo D, Navas-Cortés JA, Landa BB (2020) Relationship between soil properties and banana productivity in the two main cultivation areas in Venezuela. J Soil Sci Plant 20:2512–2524. https://doi.org/10.1007/s42729-020-00317-8

Olivares BO, Rey JC, Perichi G, Lobo D (2022) Relationship of microbial activity with soil properties in banana plantations in Venezuela. Sustainability 14:13531. https://doi.org/10.3390/su142013531

Pattison T, Smith L, Moody P, Armour J, Badcock K, Cobon J, Rasiah V, Lindsay S, Gulino L (2005) Banana root and soil health project - Australia. In: DW, Rosales FE (eds) Banana root system: towards a better understanding for its productive management. Proceedings of an international symposium international network for the improvement of banana and planta. Montpellier, France, pp 149-165

Pulido-Moncada MA, Lobo-Luján D, Lozano-Pérez Z (2009) Asociación entre indicadores de estabilidad estructural y la materia orgánica en suelos agrícolas de Venezuela. Agrociencia 43(3):221–230

Rey JC, Chacín M, Sapuky M, Núñez M, Martínez G, Rodríguez G, Espinoza J, Arturo M, Pocasangre L, Delgado E, Rosales F (2006) Land suitability for banana in Venezuelan soils and its relation to productivity. XVII Reunión Internacional ACORBAT: Banano un negocio sustentable. Joinville. Santa Catarina, Brasil. Nov 15–20. p. 362

Rey JC, Martínez G, Lobo D, Trejos J, Pocasangre L, Rosales F (2009) Aspectos sobre calidad y salud de suelos bananeros en Venezuela. Producción Agropecuaria 2:52–55. (In Spanish)

Rodríguez G, Lobo D (2004) Desarrollo y distribución de raíces en tres clones de musáceas y su relación con las propiedades de un suelo lacustrino de la Cuenca del Lago de Valencia. Rev Fac Agron (LUZ) 21(1):121–128

Rodríguez V, Malavolta E, Sánchez A, Lavoranti O (2004) Reference nutritional balance of soils and leaves in the Horn plantain crop. Bioagro 16:39–46

Rodríguez V, Silva A, Rodríguez O (2005) Nutritional balance and number of leaves as variables for yield prediction of the Horn plantain. Pesqui Agropecu Bras 40:175–177. https://doi.org/10.1590/S0100-204X2005000200011

Rodríguez G, Núñez MC, Lobo D, Martínez G, Rey JC, Espinoza J, Muñoz N, González D, Rosales F, Pocasangre L, Delgado E (2006) Banana root health in lots with different productivity levels in a soil at the oriental coast of Maracaibo Lake, Venezuela. XVII Reunión Internacional ACORBAT: Banano un negocio sustentable. Joinville, Santa Catarina, Brasil. Nov 15–20. p. 355

Salas A, Achinelly MF (2020) Community structure of soil nematodes associated with the rhizosphere of *Solanum Lycopersicum* in a major production area in Argentina: a case study among agroecosystem types. J Soil Sci Plant Nutr 20:43–54. https://doi.org/10.1007/s42729-019-00099-8

Segura RA, Serrano E, Pocasangre L, Acuna O, Bertsch F, Stoorvogel JJ, Sandoval JA (2015) Chemical and microbiological interactions between soils and roots in commercial banana plantations (*Musa* AAA. cv. Cavendish). Sci Hortic 197:66–71. https://doi.org/10.1016/j.scienta.2015.10.028

Sumner ME (1990) Advances in the use and application of plant analysis. Comm Soil Sci Plant Anal 21:1409–1430. https://doi.org/10.1080/00103629009368313

Trujillo-Narcía A, Rivera-Cruz MC, Magaña-Aquino M (2019) The burning of sugarcane plantation in the tropics modifies the microbial and enzymatic processes in soil and rhizosphere. J Soil Sci Plant Nutr 19:906–919. https://doi.org/10.1007/s42729-019-00089-w

Vega-Ávila A, Medina E, Paroldi H, Toro M, Baigori M, Vázquez F (2018) Bioindicators of soil quality of open shrubland and vineyards. J Soil Sci Plant Nutr 18:1065–1079. https://doi.org/10.4067/S0718-95162018005003002

Villarreal-Núñez J, Pla-Sentis I, Agudo-Martínez L, Villaláz-Perez J, Rosales F, Pocasangre L (2013) Soil quality indexes in areas cultivated with banana in Panama. Agron Mesoam 24:301–315. https://doi.org/10.15517/am.v24i2.12530

Viloria JA, Pineda C, López D, Gabaldón O (2003) Land suitability of the Taguaiguay irrigation system (Aragua-Venezuela) for sugarcane and banana. Agron Trop 53:189–208

Chapter 6
Potential Use of Soil Morphological Properties to Differentiate Levels of Banana Productivity

Abstract The morphological properties of the field such as texture, consistence, and structure provide a valuable tool for the study of crop productivity and its relationship with the soil, mainly due to the ease with which they can be obtained in the field. However, these morphological data are categorical variables (nominal or ordinal), measuring on scales where it is difficult to establish the origin of the property assessment scale. In this chapter, the categorical regression was applied to select soil morphological properties at sites with different levels of banana productivity in two of the main productive areas of Venezuela. For this, an optimal scaling procedure in which the morphological variables are transformed into a numerical scale can thus be entered in a categorical principal component analysis and categorical regression analysis. The categorical regression model was developed from data from six plantations growing "Gran Nain" bananas, each with two productivity levels (high and low) (see Chap. 2). Sixty-three A horizons in thirty-six soils were described using 15 field morphological variables. The optimum model selected included biological activity, texture, dry consistence, reaction to HCl, and structure type variables. These variables explained the PI with an R^2 of 0.599, an expected prediction error (EPE) of 0.645 and a standard error (SE) of 0.135 using bootstrapping, and an EPE of 0.662 with a SE of 0.236 using tenfold cross validation. These variables showed a differentiation between high and low levels of productivity. The development of healthy soil is an aspiration sought to address these limitations, using indicators based on easily measured soil morphological parameters. In the future, the categorical regression application to morphological properties of the field soil, as was done in this study, will be a tool useful for comparing suitable banana productivity sites, obtained simply and economically.

Keywords Biological activity · Qualitative soil indicators · Dry consistence · Soil structure · Texture

B. O. Olivares Campos, *Banana Production in Venezuela*, The Latin American Studies Book Series, https://doi.org/10.1007/978-3-031-34475-6_6

6.1 Introduction

Identification of the most suitable areas for banana cultivation is essential to increasing its productivity in tropical regions. Therefore, soil properties must be characterized to understand their relationship to crop productivity (Villarreal-Núñez et al. 2013; Delgado et al. 2010b). Soil morphological field properties have been recognized as a valuable tool for studying a broad range of soil characteristics, including those related to soil development in agricultural areas, for the ease and speed with which they can be described (Soil Survey Staff 2017; Calero et al. 2008; Pulido-Moncada et al. 2017).

Obtaining morphological properties of soil in edaphological studies are obtained easily and without high costs, which are found in different soil databases without being used for analytical purposes, mainly because these types of properties are categorical (nominal or ordinal) with rating scales of origins that are difficult to establish (Delgado et al. 2010a; Vaughan and Ormerod 2005; Meulman et al. 2002). Similarly, numerical relationships of different categories are still poorly developed. Such data cannot be subjected to rigorous treatment by statistical methods, such as factor analysis or multiple regression, like numerical quantitative data (Andrews et al. 2004; Linting et al. 2007; Calero et al. 2005). Therefore, there is little background for calculating soil quality indices from categorical indicators established by decision rules based on existing knowledge (Pulido-Moncada et al. 2014; Calero et al. 2018).

Soil morphology considers a set of attributes observable in the field, both on the surface and in the soil profile, which allow us to reconstruct the edaphogenetic processes and the environmental conditions and, in many cases, interpret and predict the behavior of the plants and the response of the soil against the introduction of management or changes in land use (FAO 2009). Observable attributes generally described include composition; shape; soil structure; soil organization; color and mottles; root distribution; pores; evidence of translocated materials such as carbonates, iron, manganese, carbon, and clay; and soil consistency, among others (Casanova 2005; Bravo et al. 2017).

Most of the methodologies that are currently used to characterize the fertility of soils in production and their suitability for different uses are from several decades ago. Today, these measurements have been technical with the use of sensors and various data analysis software that allow us to manage more and better information (Calero et al. 2018; Vasu et al. 2021; Olivares et al. 2022). However, much of these technologies are based on the same principles that were followed by soil science researchers in the last century, such as the case of the visual soil examination and evaluation (VSEE), which constitutes a paradigmatic case due to the simplicity of its requirements for the study of the soil. It consists of a morphological description of the structure of tilled soils, based on the visual observation of a profile (Shepherd 2009; Ball et al. 2007; Guimarães et al. 2017).

There is a wide range of studies on banana soils in Venezuela between 1970 and 2022, the most recent being those of Hernández et al. (2007), Martínez et al.

(2008), Delgado et al. (2010a, 2010b), González-Pedraza et al. (2014), Olivares et al. (2020), González García et al. (2021a, b, c), and González-Pedraza and Escalante (2021). All these studies emphasize the relationship between the reduction in productivity and the loss of soil quality due to the intensification of production systems.

Most of the studies mentioned above are based on numerical properties, while the potential of morphological variables has not yet been fully explored. Currently, categorical regression (CATREG) has shown interesting potential for the analysis of categorical data in areas such as marketing, education, and economics, among others (van der Kooij 2007; Meulman et al. 2019; Sevinç et al. 2019). However, as far as is known, it has not been widely used in agronomic sciences and never in banana fields. This is a novel element in this chapter, in which soil morphological properties are identified as potentially promising soil indicators to analyze banana productivity in Venezuelan soils.

The results of this chapter have a novel and pioneering character in the application of CATREG, as an optimal scaling procedure, which can be easily applied to other crops in other parts of Venezuela, mainly due to scientific logic that is easily transferable to other areas, beyond agricultural and soil sciences.

The studies by MacEwan and Fitzpatrick (1996), Lal (1998), and Vasu et al. (2021) state that morphological properties can improve soil evaluation, usefully taking advantage of a large amount of categorical data in universities and research institutes.

The objective of this chapter was to validate the hypothesis that the morphological properties of the soil are capable of discriminating banana productivity levels on farms in Venezuela through a categorical regression prediction model. In this case, the qualitatively estimated morphological properties of the soil could be used to improve the evaluation of banana productivity.

6.2 Results

6.2.1 Morphological Soil Properties

Tables 6.1, 6.2, 6.3 and 6.4 show the values of morphological soil properties: texture, color, structure, consistence, stickiness, plasticity, biological activity, roots, and reaction to HCl of representative profiles on the high- and low-productivity banana sites. The characteristics of high-productivity (Mollic Ustifluvents) and low-productivity (Fluventic Haplustolls) sites on Plantation PL are similar, except for texture: High-productivity soils are silty loam (Table 6.1), while low-productivity soils are generally silty clay loam (Table 6.3), with the same very dark grayish brown color (2.5 Y 3/2), and a blocky subangular structure, with moderate-to-medium particle size. At both sites, vegetation growth is limited by the high calcium carbonate content from its parent material, which originated in Lake Valencia. At

Table 6.1 Values of morphological soil properties in the A horizons of a representative profile of each high-productivity level of plantations: texture, color, and structure (Olivares et al. 2022)

Plantations	Horizon[a]	Depth (cm)	Texture	Color	Structure (type/size/grade)
PL	Ap	0–20	sil	2,5 Y 3/2	sbk/m/gm
SM	Ap	0–17	sil	2,5 Y 4/2	sbk/f/gw
PZ	Ap	0–22	ls	2,5 Y 3/2	sbk/f/gw
PZ	A1	22–44	ls	2,5 Y 4/2	sbk/f/gw
CH	Ap	0–18	sil	2,5 Y 3/2	abk/m/gw
CH	A1	18–38	sil	2,5 Y 4/2	abk/m/gm
BA	Ap	0–18	sic	2,5 Y 4/2	abk/m/gm
BA	AC	18–40	sil	2,5 Y 4/4	m/ns/ng
KA	Ap	0–24	sicl	2,5 Y 3/2	sbk/m/gm
KA	A/C	24–42	sicl	2,5 Y 4/2	m/ns/ng

Abbreviations: **texture:** *ls* sandy loam, *cl* clay loam, *s* sand, *L* loam, *lvfs* very fine sandy loam, *sic* silty clay, *sicl* silty clay loam, *sil* silty loam, *sc* sandy clay. **Structure: type:** *m* massive, *gr* granular, *sbk* subangular blocky, *abk* angular blocky, *pr* prismatic. **Size:** *ns* structureless, *vf* very fine, *f* fine, *m* medium, *c* coarse. **Grade:** *ng* structureless, *gw* weak, *gm* moderate, *gs* strong
[a]Soil Survey Staff (2014)

Table 6.2 Values of morphological soil properties in the A horizons of a representative profile of each high-productivity level of plantations: consistence when moist and dry, stickiness and plasticity, biological activity, root abundance, root size, and reaction to HCl (Olivares et al. 2022)

			Consistence							
Plantations	Horizon[a]	Depth (cm)	When dry	When moist	Stickiness	Plasticity	Biological activity	Root abundance	Root size	Reaction to HCl
PL	Ap	0–20	dsh	mfr	wss	wps	ah	raf	rsm	rhs
SM	Ap	0–17	dsh	mvfr	wss	wps	ah	ra	rsm	rhm
PZ	Ap	0–22	dsh	mvfr	wso	wpo	ah	ram	rsvf	rhno
PZ	A1	22–44	ds	mvfr	wss	wpo	am	raf	rsvf	rhno
CH	Ap	0–18	dsh	mfr	wss	wps	ah	ram	rsf	rhno
CH	A1	18–38	dsh	mfr	wss	wps	am	raf	rsm	rhw
BA	Ap	0–18	dsh	mfi	ws	wp	ah	ram	rsm	rhno
BA	AC	18–40	ds	mfr	wss	wps	am	raf	rsm	rhno
KA	Ap	0–24	dsh	mfi	ws	wp	ah	ram	rsm	rhw
KA	A/C	24–42	ds	mfr	ws	wp	am	raf	rsf	rhw

Abbreviations: **consistence: when dry:** *dl* loose, *ds* soft, *dsh* slightly hard, *dh* hard. **When moist:** *mvfr* very friable, *mfr* friable, *mfi* firm. **Stickiness (consistence when wet):** *wso* nonsticky, *wss* slightly sticky, *ws* stick. **Plasticity (consistence when wet):** *wpo* nonplastic, *wps* slightly plastic, *wp* plastic. **Root abundance**: *raf* few roots, *ram* many roots. **Root size**: *rsvf* very fine, *rsf* fine, *rsm* medium. **Biological activity**: *ah* high activity, *am* medium activity. **Reaction to HCl**: *rhno* no reaction, *rhw* weak reaction, *rhm* moderate reaction, *rhs* strong reaction
[a]Soil Survey Staff (2014)

Table 6.3 Values of morphological soil properties in the A horizons of a representative profile of each low-productivity level of plantations: texture, color and structure, consistence when moist and dry, stickiness, and plasticity (Olivares et al. 2022)

Site	Horizon[a]	Depth (cm)	Texture	Color	Structure (type/size/grade)
PL	Ap	0–28	sicl	2,5 Y 3/2	sbk/m/gm
SM	Ap	0–18	sicl	2,5 Y 3/2	abk/m/gm
SM	A1	18–44	sicl	2,5 Y 4/2	sbk/m/gm
PZ	Ap	0–18	sicl	2,5 Y 3/2	abk/m/gm
CH	Ap	0–20	sil	2,5 Y 3/2	sbk/m/gw
CH	A1	20–42	sil	2,5 Y 4/2	abk/m/gw
BA	Ap	0–14	sicl	2,5 Y 4/2	abk/m/gm
KA	Ap	0–22	sicl	2,5 Y 4/2	sbk/m/gs
KA	A/C	22–42	sicl	2,5 Y 4/4	sbk/f/gm

Abbreviations: **texture:** *Ls* sandy loam, *cl* = clay loam, *s* sand, *L* loam, *lvfs* very fine sandy loam, *sic* silty clay, *sicl* silty clay loam, *sil* silty loam, *sc* sandy clay. **Structure: type:** *m* massive, *gr* granular, *sbk* subangular blocky, *abk* angular blocky, *pr* prismatic. **Size:** *ns* structureless, *vf* very fine, *f* fine, *m* medium, *c* coarse. **Grade:** *ng* structureless, *gw* weak, *gm* moderate, *gs* strong
[a]Soil Survey Staff (2014)

Table 6.4 Values of morphological soil properties in the A horizons of a representative profile of each low-productivity level of plantations: consistence when moist and dry, stickiness and plasticity, biological activity, root abundance, root size, and reaction to HCl (Olivares et al. 2022)

			Consistence							
Site	Horizon[a]	Depth (cm)	When dry	When moist	Stickiness	Plasticity	Biological activity	Root abundance	Root size	Reaction to HCl
PL	Ap	0–28	dh	mfi	wss	wps	ah	raf	rsf	rhs
SM	Ap	0–18	dsh	mfi	ws	wp	ah	raf	rsm	rhm
SM	A1	18–44	dsh	mfi	ws	wp	am	raf	rsm	rhm
PZ	Ap	0–18	dsh	mfi	wss	wps	am	raf	rsf	rhno
CH	Ap	0–20	dsh	mfr	wss	wps	am	ram	rsf	rhno
CH	A1	20–42	dsh	mfr	wss	wp	al	raf	rsm	rhno
BA	Ap	0–14	dsh	mfi	ws	wp	am	raf	rsvf	rhno
KA	Ap	0–22	dsh	mvfi	ws	wp	ah	ram	rsm	rhno
KA	A/C	22–42	dsh	mfi	ws	wp	al	raf	rsf	rhw

Abbreviations: **consistence: when dry:** *dsh* slightly hard, *dh* hard. **When moist:** *mvfr* very friable, *mfr* friable, *mfi* firm, *mvfi* very firm. **Stickiness (consistence when wet):** *wss* slightly sticky, *ws* sticky. **Plasticity (consistence when wet):** *wps* slightly plastic, *wp* plastic. **Root abundance**: *nra* no root, *raf* few, *ram* many. **Root size**: *nra* no root, *rsvf* very fine, *rsf* fine, *rsm* medium. **Biological activity**: *ah* high activity, *am* medium activity, *al* low activity. **Reaction to HCl**: *rhno* no reaction, *rhw* weak reaction, *rhm* moderate reaction, *rhs* strong reaction
[a]Soil Survey Staff (2014)

Plantation SM, high-productivity soils (Fluventic Haplustepts) usually have high biological activity and a moderate reaction to HCl, with weak structural stability and lithological discontinuity (Tables 6.1 and 6.2). However, low-productivity soils (Fluventic Haplustolls) were limited by their susceptibility to sealing and compaction due to the high proportion of silt and fine sand (100 and 200 μm) and very fine sand (50 and 100 μm) (Table 6.3).

The high-productivity soils (Fluventic Haplustolls) at Plantation PZ have geomorphological characteristics favoring high biological activity (Table 6.2). The low-productivity (Fluventic Haplustepts) sites show an abrupt textural change coupled with heavier soil texture, which is a significant limitation for banana productivity (Table 6.3). At Plantation CH, friable consistence along with weakly adhesive and plastic characteristics favor high biological activity and the common presence of fine roots at both productivity sites (Table 6.2). In the low-productivity soils, characteristics are related to the sharp textural change in depth and poor structure. Soils at both sites are classified as Fluventic Haplustolls (Table 6.3).

At Plantation BA in the state of Trujillo, the high-productivity sites have dark grayish brown (2.5 Y 4/2) silty clay soils (Oxyaquic Ustifluvents). The consistence is firm when moist, adhesive, and plastic with high biological activity, no reaction to HCl, and abundant roots (Table 6.2). On the other hand, low-productivity soils (Typic Ustifluvents) on this plantation also have certain limitations associated with poor drainage and the presence of a water table. They are generally unstructured with abrupt textural changes and few very fine roots (Table 6.4). At Plantation KA, the high- and low-productivity soils are silty clay loam, predominantly Entisols (Oxyaquic Ustifluvents) (Table 6.1), while in lower-productivity soils, the limitations are associated with poor soil structure (Table 6.3).

6.2.2 Categorical Principal Component Analysis (CATPCA)

Figure 6.1 clearly shows that the optimal number of components is four, where the "elbow" is located in any given case, that is, the number of components from which the percentage of explained variance of the model hardly increases when considering new components. As can be seen in Table 6.5, it is a good model, since the four principal components (PC) explain 74.37% of the variance and the Cronbach's α of the model was 0.98.

Principal Component 1 (PC1) shows a positive correlation with the variables of consistence and texture (Table 6.5), since the texture variable is nominal, that is, without a prior order of categories, the scaling produced by the model is analyzed in Fig 6.2a, which defines a textural gradient from coarse-loam textures (sandy, sandy loam, loam, very fine sandy loam) to fine-loam textures (silty loam, silty clay loam, clay loam, and silt clay), with very little differentiation between the latter (practically similar scores). The texture is ordered in a "logical" way and can be interpreted coherently with the increase in consistence indicated by the positive charges

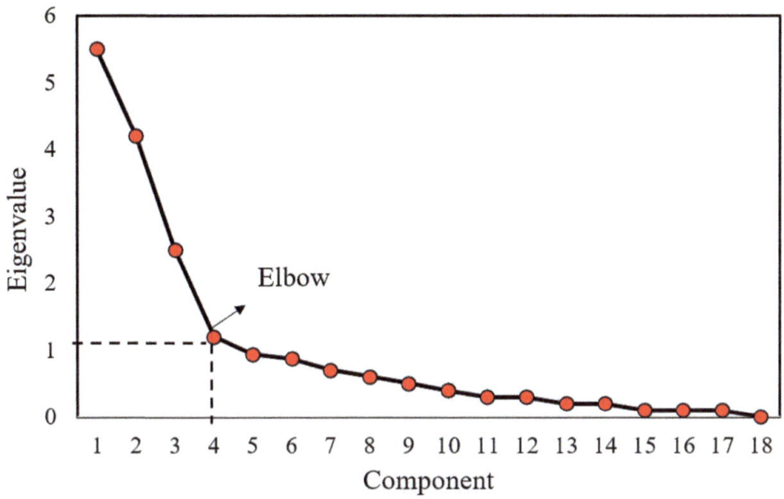

Fig. 6.1 Variations explained (eigenvalue) by the categorical principal components

of these variables in the component with good correlations ($r = 0.81$) with the optimal scalings of the consistence variables.

This PC1 is interpreted as a texture-consistence component, positively correlated with fine-textured soils and greater expression of consistence. Despite this, the component does not make it possible to significantly discriminate the production levels of the banana plantations. Table 6.6 shows the Student's t-test for the difference between means, taking PC1 and the variables correlated with it as independent variables (texture, dry consistence, wet consistence, plasticity, and adhesiveness), and the level of productivity with its associated two-level (high and low) variables as the dependent variables.

Table 6.6 shows that the component itself does not allow differentiating levels of productivity, while the variables plasticity and adhesiveness do (texture, dry consistence, and wet consistence are not significant), and then wet consistence is a good predictor of the level of productivity, in the sense that the optimal scaling means of plasticity of the high-productivity group (H) (−0.159, $N = 101$) are lower than those of the low-productivity group (L) (0.151, $N = 107$); the same occurs with stickiness (mean for $H = -0.178$ and $L = 0.167$) and, obviously, with factor scores (means for $H = -0.018$ and $L = 0.019$), since the correlation of these variables with PC1 is positive. In general, higher productivity is associated with coarse-textured soils and low-consistence "loose soils," while fine-textured, plastic, and adherent soils tend to support low banana production.

Principal Component 2 (PC2): It is positively related to biological activity traits in the horizons (both "biological activity" and root frequency and thickness) and negatively related to hue and chroma (redness of the horizon) (Table 6.5). On the other hand, the hue (nominal variable) is ordered logically, marking a gradient from yellow to reddish-brown horizons (Fig. 6.2b). It is interesting to observe in Fig. 6.2b

Table 6.5 Categorical principal component analysis of morphological soil indicators: component loadings and variance explained by principal components (PC) with varimax rotation

	Rotated component loadings				Total variance					
Variable	PC1	PC2	PC3	PC4	PC1	PC2	PC3	PC4	VAF_j	$PVAF_j$
Thickness	−0.17	−0.22	−0.09	0.06	0.02	0.04	0.01	0	0.09	0.51
Texture	**0.90**	0.12	−0.06	−0.05	0.8	0.01	0,00	0	0.83	4.62
Hue	−0.06	**−0.76**	−0.01	−0.21	0,00	0.57	0,00	0.04	0.62	3.49
Value	−0.12	0.07	−0.43	**0.67**	0.01	0,00	0.19	0.45	0.66	3.69
Chroma	−0.01	**−0.56**	−0.15	−0.15	0,00	0.31	0.02	0.02	0.36	2.04
Structure type	−0.11	0.33	**0.91**	−0.10	0.01	0.11	0.82	0.01	0.96	5.36
Structure size	−0.12	0.3	**0.91**	−0.10	0.01	0.09	0.84	0.01	0.96	5.34
Structure grade	−0.11	0.31	**0.91**	−0.10	0.01	0.09	0.84	0.01	0.96	5.35
Dry consistence	**0.86**	0.14	−0.08	0.11	0.75	0.02	0,00	0.01	0.79	4.40
Moist consistence	**0.88**	0.16	−0.14	0.12	0.77	0.02	0.02	0.01	0.83	4.60
Stickiness	**0.88**	0.07	0.02	−0.17	0.77	0,00	0,00	0.02	0.81	4.50
Plasticity	**0.88**	0.03	−0.04	−0.17	0.78	0,00	0,00	0.03	0.81	4.55
Biological activity	0.22	**0.77**	0.03	−0.17	0.04	0.6	0.11	0.03	0.8	4.47
Root abundance	0.13	**0.84**	0.25	−0.19	0.01	0.71	0.06	0.03	0.83	4.65
Root size	0.14	**0.84**	0.24	−0.17	0.02	0.7	0.06	0.03	0.81	4.53
Cutans thickness	−0.05	−0.18	0.01	**0.87**	0,00	0.03	0,00	0.77	0.8	4.48
Reaction to HCl	0.03	0.23	−0.13	**0.69**	0,00	0.05	0.01	0.48	0.56	3.13
Cutans frequency	−0.05	−0.17	0,00	**0.88**	0,00	0.3	0,00	0.78	0.82	4.55
				VAF_s	4.08	3.46	3.03	2.79	**13.38**	**74.37**
				$PVAF_s$	22.7	19.27	16.86	15.53	**74.37**	**13.38**

VAF_j = variance accounted for by the *j*th indicator. $PVAF_j$ = percentage of variance accounted for by the *j*th variable = ($VAF_j \times 100$)/18 (18 is the number of variables)
VAF_s = variance accounted for by the *s*th component. $PVAF_s$ = percentage of variance accounted for by the *s*th component = ($VAF_s \times 100$)/4 (4 is the number of components). Bold numbers: variables with loadings greater than |0.50|.

that the truly discriminating soil color is 2.5Y, which presents very different scores from the others (all the same), and, in this case, given the direction of the factorial charges, the 2.5Y horizons seem to be the most apt for biological activity.

In this case, none of the variables well correlated with the PC2, or the scores (frequency and root thickness, hue and chroma, p-value >0.05), except biological activity, allow us to significantly discriminate production levels (Table 6.6). The highest mean values of the transformed variable are associated with high productivity (means for H = 0.156 and L = −0.147) (although the correlation of biological activity, as indicated by the charge of this variable, is positive with PC2, the signs of the means of the production groups in both are opposite due to the high dispersion and p-value of PC2).

Principal Component 3 (PC3): This is a purely structural component, since it is positively related to the type, class, and grade of structure, without any other variable presenting significant correlations with it (Table 6.5). The type of structure was introduced in the analysis as a nominal variable; the ordering resulting from the categories is the one observed in Fig. 6.2c. According to this arrangement, there

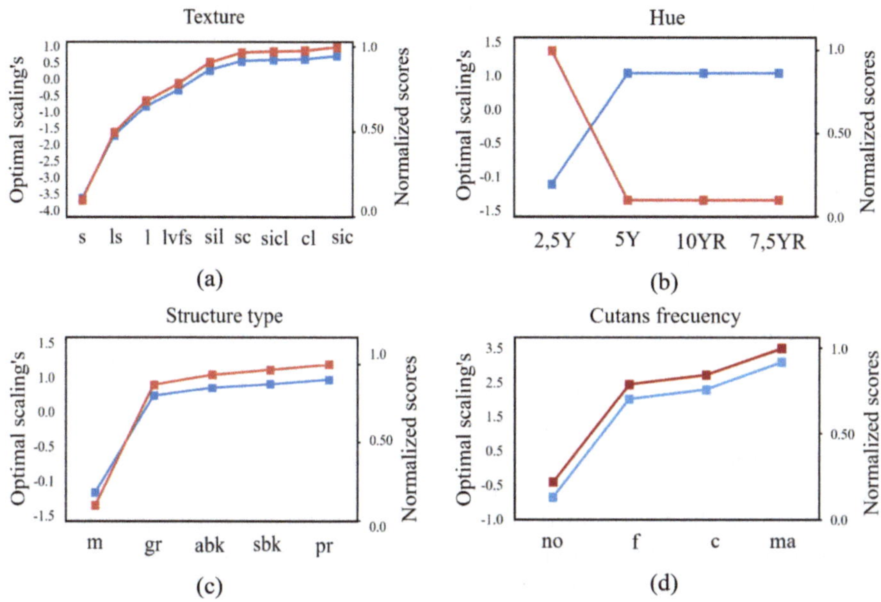

Fig. 6.2 Optimal scaling of morphological properties selected for their ordinal and nominal character: (**a**) texture, (**b**) hue moist, (**c**) structure type, and (**d**) cutans frequency. Abbreviations: texture: ls sandy loam, cl clay loam, s sand, L loam, lvfs very fine sandy loam, sic silty clay, sicl silty clay loam, sil silty loam, sc sandy clay. Structure: type: m massive, gr granular, sbk subangular blocky, abk angular blocky, pr prismatic. Frequency: no presence, f few, c common; ma many

Table 6.6 Student's *t*-test of the principal components scores and the optimal scaling of the soil morphological variables (PC)

Component	Variable	*t*	d.f.	Signification
1	PC1 (component scores)	0.25	197.96	0.79
	Stickiness (optimal scaling)	2.46	190.76	0.01
	Plasticity (optimal scaling)	2.22	190.18	0.02
2	PC2 (component scores)	1.21	206.00	0.22
	Biological activity (optimal scaling)	2.46	190.76	0.01
3	PC3 (component scores)	−2.14	169.00	0.03
	Structure type	−0.26	206.00	0.79
	Structure size	−0.99	206.00	0.92
	Structure grade	−0.00	206.00	0.99
4	PC4 (component scores)	−0.12	206.00	0.90
	HCl (optimal scaling)	−0.02	206.00	0.98
	Cutans frequency	−3.09	159.00	0.02
	Cutans thickness	−3.04	163.00	0.00

d.f. = degrees of freedom; *p*-value >0.05

seems to be a great difference between soils without structure (massive, the same for structure size and grade) and structured soils, while the type, size, and grade soil structure in non-massive soils explain little, by themselves, system variance. It is interesting to note that the means of the PC3 factorial scores do allow us to discriminate significantly between high and low production (Table 6.6).

In this way, it is verified that the loss of structure soil (lower PC3 scores) (Table 6.6) is associated with lower productivity in bananas. On the other hand, individually, none of the variables considered make it possible to differentiate between productivity levels (only the PC3 score, which indicates a high collinearity in the variables of type, size, and grade of structure).

Principal Component 4 (PC4): Positively correlates with value, reaction to carbonates (HCl), and the frequency and thickness of cutans (Fig. 6.2d). The PC4 scores do not show significantly different mean values for high- and low-productivity levels, although the variables related to the slats do (Table 6.6). Therefore, the presence of horizons with cutans (high scores, frequent cutans, and thickness) is associated with higher levels of productivity.

In summary, the transformed variables cutans frequency, cutans thickness, biological activity, plasticity and adhesiveness, and the factorial scores of PC3 (component basically related to soil structure) allow to discriminate between high- and low-productivity banana soils.

On the other hand, the correlations between the transformed morphological variables and the factor scores with the quantitative productivity index (PI) are shown in Table 6.7. In the case of the PI correlations with the different variables, taking into account all the horizons of the profile ($N = 208$), the association of the variables of cutans, plasticity, and adhesiveness is maintained, as well as the structural component (PC3) and biological activity (in this case, with a slightly insignificant p-value, 0.058), with an indicator of banana productivity (in this case, the PI, not the H and L levels); although the r are very low (<0.200), this could be expected, because the PI is calculated by profile, not by horizon. For this reason, the information of the transformed variables regarding the productivity of the banana trees taking into account the groups or the PI is consistent.

In the case of the PI correlations with the average weighted by the thickness of the horizon for 60 cm ($N = 36$), it is obtained that only the correlations with the values averaged to the total of the profile are even worse; it is not shown in the Table 6.7.

When comparing the correlation at the profile level, the cutans present notable correlations (although not significant, but with relatively low values of p-value $\approx$ 0.07) and the biological activity, maintaining to a certain extent the information of the horizons. On the other hand, a significant and positive correlation appears between PI and root abundance and root frequency (p-value = 0.059, almost significant), which was not previously detected in the horizons, while the correlations with plasticity and adhesiveness are lost. Finally, what is most notable is that the PC with a significant correlation with PI is PC4 (the "cutans" component), instead of PC3 (the "structural" component), as indicated above for all horizons and in line with the ability to discriminate H and L groups.

Table 6.7 Correlation of the results according to the PI with soil horizons ($N = 208$) and PI with soil profiles ($N = 63$)

PI with soil horizons (N = 208)	*r*	*p*
CP1	−0.014	0.839
CP2	−0.097	0.165
CP3	**0.129**	**0.063**
CP4	0.063	0.363
Stickiness	**−0.166**	**0.016**
Plasticity	**−0.156**	**0.024**
Cutans thickness	**0.176**	**0.011**
Cutans frequency	**0.184**	**0.008**
Biological activity	**0.132**	**0.058**
PI with soil profiles (N = 63)	*r*	*p*
CP1	−0.037	0.829
CP2	−0.064	0.710
CP3	0.152	0.378
CP4	**0.394**	**0.017**
Biological activity	**0.459**	**0.005**
Root abundance	**0.361**	**0.031**
Root frequency	**0.306**	**0.059**
Cutans thickness	**0.302**	**0.073**
Cutans frequency	**0.304**	**0.071**

p-value >0.05 (numbers in bold: significant)

The variables outlined in the Table 6.7 also discriminate significantly between the groups of high and low productivity (including PC4, although with p-value = 0.089, $N = 36$), so the results in discrimination between H and L profile levels are similar to the results at the level of the total horizons.

6.2.3 Model Selection

From the original set of 15 predictors, we obtained a significant model with a high coefficient of determination, $R^2 = 0.746$ (p-value <0.007). However, as most of the regression coefficients in that first model were not significant (Table 6.8), they had to be refined. The coefficients of texture, structure type, dry consistence, biological activity, and reaction to HCl were significant (p-value <0.05). These variables should be considered in a stepwise procedure since they respond to soil processes related to plant physiology and eventual yield. Therefore, to fit a better model with significant coefficients while keeping R^2 as high as possible, we checked the MSE as a measure of the expected prediction error (EPE) given by alternative models in which predictors were progressively removed (Fig. 6.3). Thus, the optimum model included five predictors (Fig. 6.3). Figure 6.4 shows the expected prediction errors in the nonregularized and regularized models. Considering the average EPEs of both bootstrap and cross-validation estimations, the most accurate models seem to

Table 6.8 Regression coefficients, correlation with the outcome, and tolerance of the model from the initial set of 15 predictors (Olivares et al. 2022)

	Coefficients					Correlation with the outcome		Tolerance	
Predictors	Beta	B[a]	gl	*F*	*p*-value	*r*	Partial *r*	After	Before
Texture	0.397	0.224	5	3.127	**0.023**	0.118	0.556	0.723	0.704
Moist hue	0.193	0.195	2	0.980	0.388	−0.017	0.275	0.557	0.617
Moist value	0.175	0.203	1	0.742	0.396	0.049	0.272	0.663	0.641
Moist chroma	0.245	0.277	2	0.780	0.468	−0.092	0.298	0.414	0.411
Structure type	0.409	0.216	3	3.598	**0.026**	0.275	0.575	0.748	0.263
Structure size	−0.005	0.397	1	0.000	0.989	−0.251	−0.007	0.397	0.148
Structure grade	0.281	0.547	1	0.264	0.612	−0.298	0.299	0.316	0.151
Dry consistence	−0.612	0.234	2	6.829	**0.004**	−0.297	−0.645	0.483	0.628
Moist consistence	−0.515	0.349	3	2.178	0.113	−0.253	−0.508	0.334	0.312
Stickiness	−0.609	0.462	2	1.735	0.195	−0.193	−0.469	0.193	0.189
Plasticity	0.430	0.383	2	1.256	0.300	−0.186	0.334	0.172	0.189
Biological activity	0.792	0.311	2	6.480	**0.005**	0.273	0.688	0.364	0.221
Root abundance	−0.112	0.452	1	0.061	0.807	0.172	−0.137	0.392	0.192
Root size	0.052	0.409	3	0.016	0.997	0.119	0.072	0.493	0.370
Reaction to HCl	−0.532	0.240	2	4.898	**0.015**	−0.252	−0.645	0.638	0.398

[a]Bootstrap SE estimates of beta coefficients (1000 bootstrap resamplings). Numbers in bold: significant

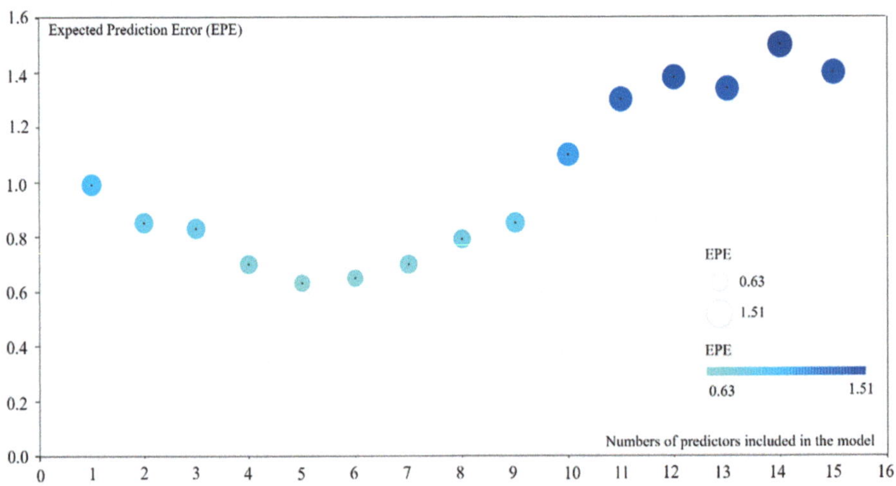

Fig. 6.3 Expected prediction error (EPE) of 15 different nonregularized models, employing the 0.630 bootstrapping procedure

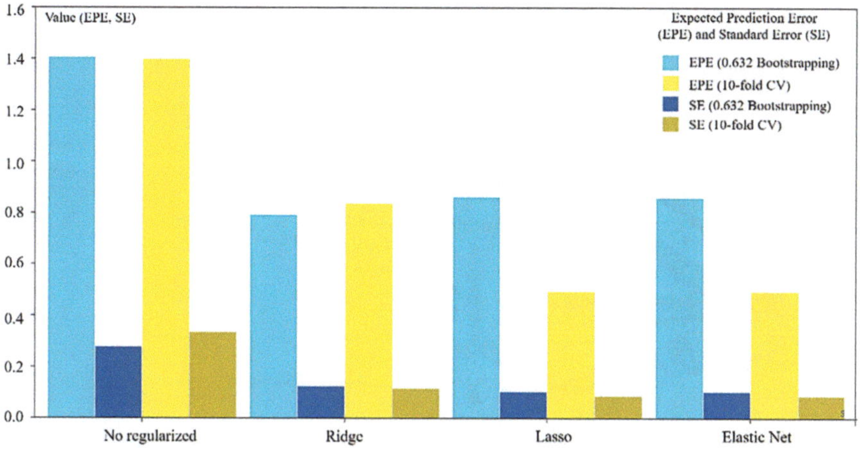

Fig. 6.4 Expected prediction error (EPE) and standard error (SE) of regularized CATREG models for the initial set of 15 predictors

be elastic net (average EPE of 0.676) and Lasso (average EPE of 0.677), because they had the lowest EPE. We used the Lasso model to select the optimal set of predictors since it is simpler and easier to interpret, and in addition, there are no appreciable differences in accuracy from the elastic net model.

The Lasso paths (Fig. 6.5) drawn represent the 42 regularized models performed by increasing λ by 0.02 per step, from a standardized sum of coefficients of 1 (unshrunk coefficients, right side of the graph) to 0 (left side of the graph). The variables with the regression coefficients that are shrunken earliest, lowering their penalized coefficients to zero, should be taken as those with the lowest predictive power, regardless of their starting value (right of the plot), so they can be removed from the model. On the contrary, coefficients that keep nonzero values at 1-SE of the regularized optimal model may be left in the theoretical model. Our optimum model included eight predictors with an λ of 0.260, an EPE of 0.789, and a SE of 0.122, while the most parsimonious model, with an EPE of 0.861 (λ of 0.400), was fitted using seven predictors. In this case, there was no important difference in simplicity (parsimony) between the two regularized models. Only the standardized coefficient of root abundance was reduced to zero in the most parsimonious model concerning to the optimum. The eight predictors for the optimum model were (1) biological activity, (2) dry consistence, (3) texture, (4) structure grade, (5) structure type, (6) HCl reaction, (7) stickiness, and (8) root abundance. From this starting set of predictors (Model 4 in Fig. 6.6), alternative nonregularized models of seven (Model 3), six (Model 2), and five (Model 1) predictors were tested in a backward stepwise regression to find the statistically significant model with the highest R^2 and minimum prediction error (Fig. 6.6). Model 1 was finally selected, because it met all these conditions: statistical significance of R^2 (p-value <0.0001) and of all its coefficients (p-value <0.050) while yielding the minimum bootstrapped EPE (0.645;

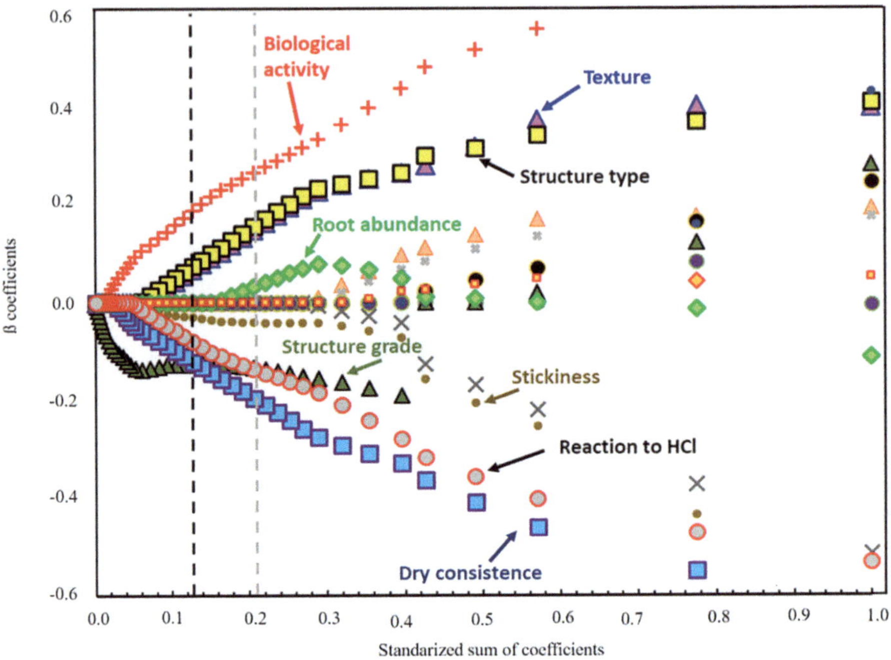

Fig. 6.5 Categorical regression (CATREG)-Lasso coefficients path estimated with the 0.632 bootstrap (dashed bar shows the most parsimonious (black vertical line) and the optimal (grey vertical line) models within one standard error). The coefficients in the right side of the plot (standardized sum of coefficients of 1.000) are the same of that in Table 6.5 (Olivares et al. 2022)

see also Fig. 6.3) and a good enough R^2 (0.599) (Fig. 6.7) close to the model with the highest determination coefficient (Model 4, $R^2 = 0.645$).

The statistical significance of the regression coefficients was the main issue with these alternative models, because they all yielded quite similar prediction errors (within one SE) and significant overall fit (p-value <0.0001). In Model 4, with eight predictors, elimination of root abundance was supported by the *1-SE* rule, as discussed above, while stickiness and structure grade presented coefficients far from statistical significance (p-values = 0.195 and 0.612, respectively) in the starting model (Table 6.8), as well as lower tolerances of all predictors (except plasticity). Therefore, Model 1 with five predictors seemed to be the best model for predicting the PI, and it was chosen as the definitive CATREG optimal scaling model (Fig. 6.7).

According to the sign of the coefficients, the ordinal variable, biological activity, is positively correlated with the dependent variable, PI, which implies that higher biological activity is correlated with higher productivity, while harder (drier consistence) soils with carbonates (reaction to HCl) are associated with lower productivity. The transformation functions (optimal scaling) indicate the linearity of this relationship, as well as the direction of the variation of the nominal variables, texture, and structure (Fig. 6.8).

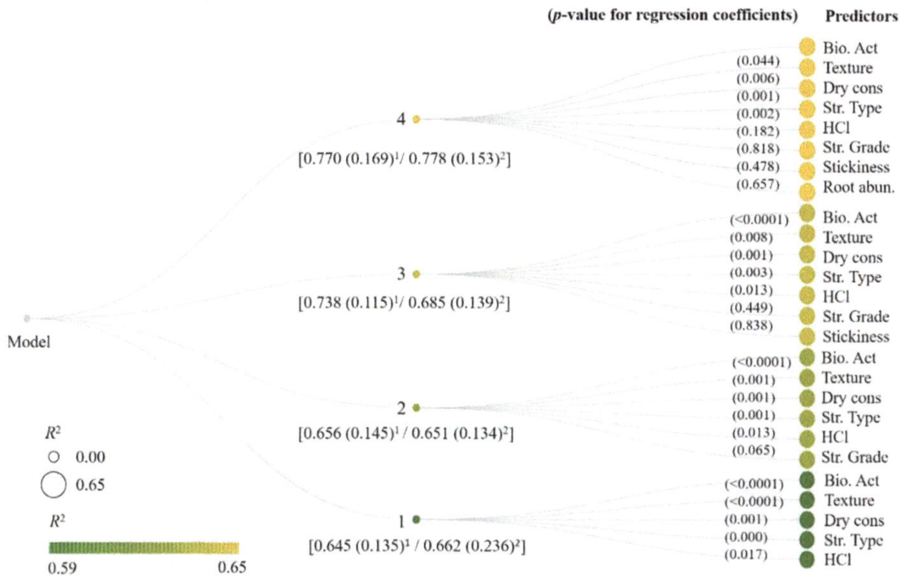

Fig. 6.6 Selection of the optimal CATREG model, EPE: expected prediction error; SE: standard error (in bracket) of bootstrapped and tenfold cross validation, Predictors (*p*-value for regression coefficients). [1]EPE and SE (in bracket) of bootstrapped cross validation. [2]EPE and SE (in bracket) of tenfold cross validation. Abbreviations: Bio. Act. biological activity, Str. Grade structure grade, Str. Type structure type, dry cons. dry consistence, HCl reaction to HCl

6.2.4 Optimal Scaling of the Field Soil Morphological Variables

Figure 6.8 shows optimal scaling of the selected categorical regression model. Of the transformation functions, dry consistence and biological activity were practically linear, which implies a proportional increase (biological activity) (Fig. 6.8d) or decrease (dry consistence) (Fig. 6.8b) in productivity index (PI) with these variables. Two soil productivity groups were formed according to the acid reaction to their carbonate content: (1) no reaction (rhno) and weak reaction (rhwo) and (2) moderate (rhm) or strong (rhs) reaction (Fig. 6.8e). As this predictor has a negative correlation (beta coefficient) with PI, moderate or strong reactions must be associated with a decrease in productivity.

For structure type (Fig. 6.8c), categories were ordered (entered in the model as nominal and, therefore, with unrestricted order) as prismatic (pr), subangular blocks (sbk), massive (m), and angular blocks (abk). As the beta coefficient was positive, a positive correlation was defined between more developed (angular) structures and productivity, while prismatic structure, "extreme" development of the angular structure, was as unfavorable as massive structure, which could be explained by its characteristic loss of porosity (macroporosity). Texture categories, also entered as nominal, were ordered on a rough gradient from finer clay (cl) to loamy textures (l),

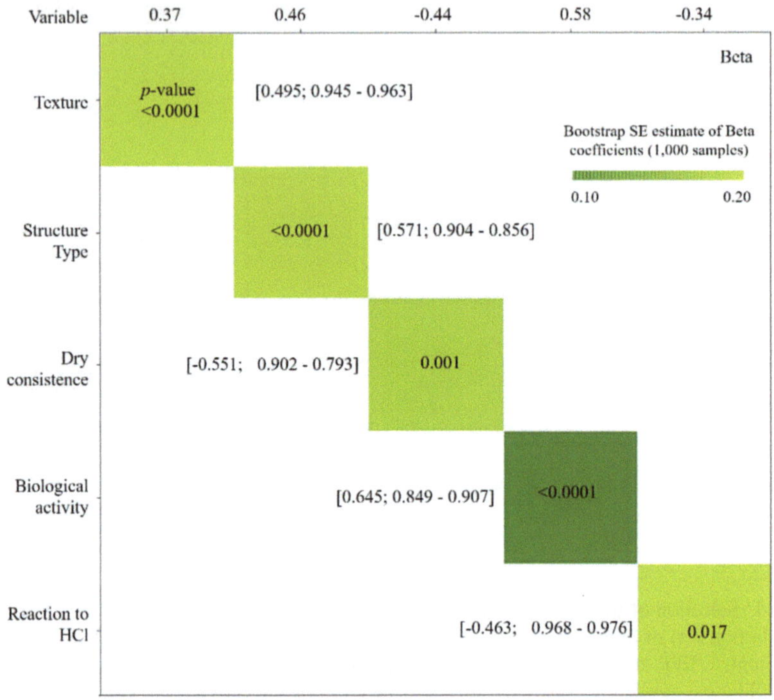

Fig. 6.7 Bootstrap coefficients (beta) in *y* axis, *p*-value inside the box, and regression coefficients: correlation with the outcome [partial *r* and tolerance after and before] of the optimal CATREG model. Bootstrap SE estimate of beta coefficients (1000 samples)

and as the beta coefficient was positive, this was the most closely correlated with productivity index (PI) (Fig. 6.8a).

6.2.5 Prediction Accuracy

Figure 6.9 shows the prediction plot of observations versus predictions. As mentioned above, the CATREG model with the lowest bootstrap estimate of the prediction error was selected. However, it yielded a relatively high MSE (EPE) of 0.645 (RMSE of 0.803), accounting for 11.7% of the outcome ranking, which varies from −4.178 to 2.680 (*x* axis in Fig. 6.9).

For a more general assessment of model performance, the observed and predicted dependent variables were classified into two groups or classes, with a PI of zero as the cutoff point: one level, *N*, with PI <0 (low PI) and another with PI >0 (high PI). It should be mentioned that most of the horizons of soils in high-productivity sites also had positive predicted PI values. Table 6.9 shows the

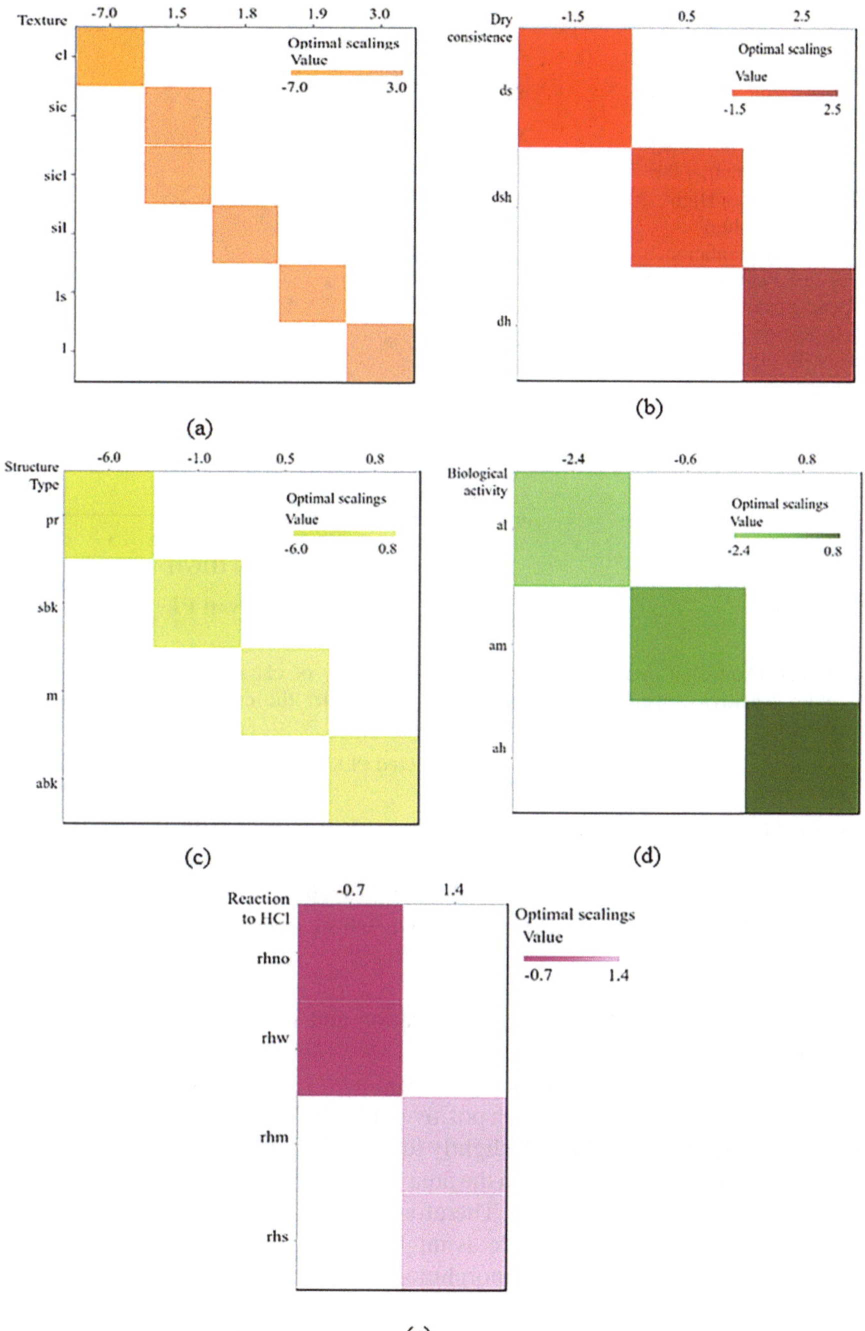

Fig. 6.8 Optimal scalings of (**a**) texture, (**b**) dry consistence, (**c**) structure type, (**d**) biological activity, and (**e**) reaction to HCl. Abbreviations: texture: ls sandy loam, cl clay loam, l loam, sic silty clay, sicl silty clay loam, sil silty loam. Dry consistence: ds soft, dsh slightly hard, dh hard. Structure type: m massive, pr prismatic, sbk subangular blocky, abk angular blocky. Biological activity: al low activity, am medium activity, ah high activity. Reaction to HCl: rhno no reaction, rhw weak reaction, rhm moderate reaction, rhs strong reaction)

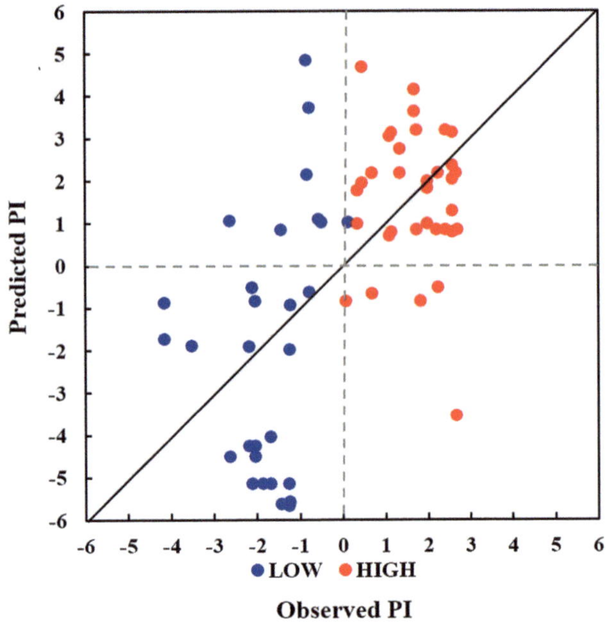

Fig. 6.9 Observed vs. predicted values of the productivity index (PI) for the five predictors optimal categorical regression model. The color of each point shows whether the plot belongs to a High (red) or low (blue) productivity plantation. Gray lines: cutoff value for dependent variable classification in P (PI >0) and N (PI <0) groups (Olivares et al. 2022)

Table 6.9 Confusion matrix and discrimination accuracy of classified values of the outcome variable: sensitivity, specificity, efficiency, and area under the curve ROC (AUC) (Olivares et al. 2022)

		Predicted PI level		
		P	N	Total observed
Observed PI level	P	30	5	35
	N	7	21	28
Total predicted		37	26	63

P = positive PI (>0), N = negative PI (<0). Sensitivity = 0.86, specificity = 0.75, efficiency = 0.81, AUC = 0.834

confusion matrix of the resulting classification and various measures of their discrimination accuracy.

Under these conditions, the model showed high sensitivity of 86% and misclassified only five cases recognized as positive or high PI (bottom right-hand quadrant in Fig. 6.9). Specificity, although slightly lower (75%), was also satisfactory, as was classification efficiency (81%) and the area under the ROC curve (AUC), which was 0.834, being considered excellent. Therefore, although the prediction based on continuous PI values is not as reliable as might be desired, prediction of the level of productivity of the plot from the morphological indicators would be quite suitable. Finally, as deduced from Fig. 6.4, the statistical relationship between predicted PI and a priori plot type, high or low productivity, was also significant: The mean predicted PI for high-productivity was 1.579 (SD = 3.018), and mean predicted PI for low productivity was −1.872 (SD = 1.642), and t = −5.530 (*p*-value = 0.000).

6.3 Discussion

The results of this research show that it may be feasible to establish a quantitatively interesting relationship between banana productivity and key morphological properties of the soil. The determination of a set of five variables that best explain banana productivity levels in six sites in Venezuela was possible by using categorical regression with optimal scaling, obtaining nonlinear transformations (Lasso) to generate a sparse model with stable predictors and using bootstrap 0.632 for the evaluation of the accuracy of the prediction.

The morphological variables identified were texture, dry consistence, type of soil structure, indicators of biological activity, and HCl reaction. The results of van der Kooij (2007) are similar to the prediction accuracy of this model. The applied categorical regression procedure adequately discriminated areas with high- and low-productivity soils on the same farm and characterized the trend in productivity variation among banana plantations as soil morphological variables changed (Fig. 6.9).

The key morphological properties of the soil reported are usually found in the descriptions of the soil profile in the field, so they can be useful for banana producers, environmental managers, and agricultural planners whose interpretation can be done easily. The results of the categorical regression indicate that texture, grade structure, type of soil structure, and biological activity are closely related to banana productivity in tropical soils.

According to Casanova (2005), sand, silt, and clay contribute to certain characteristics of the soils; therefore, it is important to know both the texture of the soil in the superficial horizon and the textural class of the soil profile, since a B horizon with high clay content can affect the air-water relationship and restrict vertical root penetration.

The model calibrated identified this, with a higher score in productivity for the loamy textures. This is in line with observations at Plantations PL, BA, and KA, where the high-productivity plots had loamy textures (L/L–SL) and low-productivity plots had a moderate to fine texture in the A horizon. In the largest plantations, the high-productivity plots were in the alluvial plains (BA, KA, SM, PZ, and CH), which had a higher clay content than the lower-productivity plots on the same plantations. A study by Vaquero (2005) identified that areas with low banana productivity had soil horizons with coarse textures (very sandy soils) and very fine textures (clay > 60%) due to direct action on water retention capacity and the permeability. The various investigations of Vaquero (2005), King et al. (2009), and Delgado et al. (2010a) conclude that medium- to fine-textured soils (loam to silt loam) with adequate structure and porosity develop a deeper and more extensive root system in bananas grown in tropical soils.

Regarding the indicator of biological activity, a positive correlation with higher productivity was evidenced. In this sense, this indicator is key for the promotion of good agricultural practices in the farms studied, due to its link between soils with good biological activity and higher banana productivity. According to Casanova

(2005), tillage operations (ploughing), fertilizer applications, the use of manure and liming of soils, and the use of fungicides and herbicides represent agricultural practices that usually reduce the microbial population in soils or favor the predominance of a certain class of biological organisms over another.

The type of soil structure (Figs. 6.7 and 6.8) was also key in the reduction of banana productivity levels, this being key morphological property in the process of soil formation and anthropogenesis (Hernández et al. 2010). The report by Voorhees et al. (1971) establishes that the soil structure is a fundamental function in the pedogenesis and nutrition of plants. On the other hand, the study by Casanova (2005) shows that granular and porous structures are the most desirable for plant growth, since the space between aggregates allows air to circulate and water to be retained inside the soil aggregate.

Optimum scaling analysis, identifying the association between soils and massive or highly developed prismatic structures had a lower productivity score. The research by Gauggel et al. (2005) demonstrated a rapid deterioration of the banana root system in thick and very thick blocks and prismatic structures, this characteristic coinciding with low-productivity soils in the plantations located in the state of Trujillo (BA and KA). Likewise, the studies by Gauggel et al. (2005) and Villarreal-Núñez et al. (2013) noted the deterioration of the root system in those soils with massive clayey structures at shallow depths.

The investigations of Johannes et al. (2017) and da Silva et al. (2014) show that the soil structure is considered an indicator of soil quality that directly affects productivity. Technically, the "best" soil structure corresponds to a block or granular (crumb) structure, because it favors adequate movement of air and water in the soil and has a lower bulk density, and, therefore, a higher porosity. On the contrary, soils with a poor structure tend to be compacted, with little aeration, less moisture retention capacity, and greater susceptibility to water erosion.

Our results indicated that visual evaluation of soil structure is a useful and reliable semiquantitative method that integrates physical functions (e.g., aeration, root growth, and water availability) related to the structural and physical quality of soils. Therefore, that visual evaluation could be used as an alternative or complementary tool to assess the impacts of cropping practices on soil structural quality. However, it is necessary to consider that the quantitative physical properties of the soil have site-specific responses, which are highly influenced by the inherent characteristics of the soil and, therefore, become one of the drawbacks of using one of these properties alone as an indicator of soil structure or quality.

Dry consistence was the most significant variable associated with PI (Fig. 6.7), where consistence was strong on lower productivity soils (Figs. 6.6 and 6.7). A study by Dorel (1993) in alluvial, firm, and very firm soils with weak or no structure can cause compaction and decrease the productivity of bananas in tropical soils. The results of Vaquero (2005), who found that banana root density was higher in soils with a friable to very friable consistence, low bulk density, and low resistance to penetration, also coincide with ours. In the sampled areas in Trujillo and Aragua, root density was radically reduced in areas with high resistance to penetration associated with firm or very firm soil consistence and high bulk density. According to

Casanova (2005), loose and porous soils tend to have little weight per unit volume (low bulk density), and those more compact soils (little pore space) will have high bulk density.

The last indicator in the model is the high carbonate content, associated with the reaction to HCl (Figs. 6.7 and 6.8), which can potentially decrease banana productivity according to Cigales and Pérez (2011). This is associated with the limited response to fertilization of very calcareous soils, which can impair the development of the bunch, reduce the size of the pseudo-stem and the height of the plant, and sometimes facilitate the appearance of diseases (Olivares et al. 2020). Research by Vanilarasu and Balakrishnamurthy (2014) and El-Khawaga (2013) indicates that when carbonate accumulates at a certain depth in the soil profile, the apical bud can die, even after normal initial development.

Finally, it shows that through categorical regression analysis with optimal scaling, an operational model capable of incorporating the effect of qualitative soil information on banana productivity can be obtained. However, these results must be properly interpreted, in case they want to be applied in other banana areas, which could represent a useful tool for farmers, technicians, planners, and other actors to identify the best areas for banana plantations. This type of research based on relatively simple field evaluations at a moderate cost is advantageous, since it emphasizes the use of soil information from field studies that are not used for analytical purposes but for descriptive ones.

6.4 Conclusions

The soil morphological indicators obtained in the categorical regression model with optimal scaling have an interesting agronomic relationship with banana productivity in Venezuelan soils. The proposed model could be used in the field for the reliable identification of areas of high and low potential banana productivity in other banana-growing areas in Venezuela after a local assessment. These morphological properties may contribute to the long-term sustainability of banana soils in Venezuela.

The results show that the quantitative transformation of the morphological properties of the soil through the application of categorical regression has great potential for the development of future studies in Venezuela, because it is a methodology easily applicable to other crops, with little or no expertise.

Calibration of similar correlations between banana productivity indicators, or even actual yield records, and soil morphological properties in specific areas, using the methodology proposed in this study, could be done in a few months at a relatively moderate cost, which would compensate with the savings in planting the banana plots in the most suitable areas.

References

Andrews SS, Karlen DL, Cambardella CA (2004) The soil management assessment framework: a quantitative soil quality evaluation method. Soil Sci Soc Am J 68:1945–1962. https://doi.org/10.2136/sssaj2004.1945

Ball BC, Batey T, Munkholm LJ (2007) Field assessment of soil structural quality a development of the Peerlkamp test. Soil Use Manag 23:329–337

Bravo C, Torres B, Alemán R, Marín H, Durazno G, Navarrete H, Gutiérrez ET, Tapia A (2017) Indicadores morfológicos y estructurales de calidad y potencial de erosión del suelo bajo diferentes usos de la tierra en la Amazonía ecuatoriana. Anales de Geografía de La Universidad Complutense 37(2):247

Calero J, Serrano JM, Aranda V, Sánchez D, Vila MA, Delgado G (2005) Analysis and characterization of olive tree cultivation system in Granada province (South of Spain) with optimal scaling and multivariate techniques. Agrochimica 49:118–131. Available online at https://n9.cl/k3rdk. Accessed 07 July 2020

Calero J, Delgado R, Delgado G, Martin-Garcia MJ (2008) Transformation of categorical field soil morphological properties into numerical properties for the study of chronosequences. Geoderma 145:278–287. https://doi.org/10.1016/j.geoderma.2008.03.022

Calero J, Aranda V, Montejo-Raez A, Martin-Garcia JM (2018) A new soil quality index based on morpho-pedological indicators as a site-specific web service applied to olive groves in the Province of Jaen (South Spain). Comput Electron Agric 146:66–76. https://doi.org/10.1016/j.compag.2018.01.016

Casanova E (2005) Introducción a la ciencia del suelo. Consejo de Desarrollo Científico UCV: Caracas, Venezuela

Cigales M, Pérez O (2011) Soil variability and water requirement of banana cultivation in a locality in the Pacific of Mexico. Av Investig Agropecu 15:21–31. https://n9.cl/hxjtg. Accessed 07 July 2020

da Silva AP, Ball BC, Tormena CA, Giarola NF, Guimarães RM (2014) Soil structure and greenhouse gas production differences between row and interrow positions under no-tillage. Sci Agric 71:157–162

Delgado E, Rosales F, Trejos J, Villalobos M, Pocasangre L (2010a) Soil quality and health index for banana plantations in four countries of Latin America and the Caribbean. Bioagro 22:53–60. https://n9.cl/qm3e. Accessed 07 Feb 2021

Delgado E, Trejos J, Villalobos M, Martínez G, Lobo D, Rey J, Rodríguez G, Rosales F, Pocasangre L (2010b) Determination of a soil quality and health index for banana plantations in Venezuela. Interciencia 35:927–933. https://n9.cl/5ly9u. Accessed 07 Feb 2021

Dorel M (1993) Banana development in an andosol in Guadeloupe: effect of soil compaction. Fruits 48:83–88. https://n9.cl/vahwkz. Accessed 07 Feb 2021

El-Khawaga AS (2013) Response of grand Naine banana plants grown under different soil moisture levels to antitranspirants application. Asian J Crop Sci 5:238–250. https://doi.org/10.3923/ajcs.2013.238.250

FAO (2009) Organización de la Naciones unidas para la Alimentación y Seguridad Alimentaria. Guía para la descripción de los suelos, Roma, p 99

Gauggel CA, Sierra F, Arévalo G (2005) The problem of banana root deterioration and its impact on production: Latin America's experience. In: Turner DW, Rosales FE (eds) Banana root system: towards a better understanding for its productive management. Proceedings of an international symposium held in San José, Costa Rica on 3–5 November 2003. International Network for the Improvement of Banana and Plantain (INIBAP), Montpellier, pp. 13–22. https://n9.cl/b3wi8. Accessed 07 Feb 2021

González García H, González Pedraza AF, Atencio J, Soto A (2021a) Evaluation of quality of banana soils through microbial activity in the south the lake of Maracaibo, Zulia state, Venezuela. Rev Fac Agron (LUZ) 38(2):216–240. https://n9.cl/wdnr5. Accessed 19 February 2021

González García HG, Pedraza AFG, Yzquierdo GR, Pacheco RL, Vásquez MB (2021b) Vigor of plantain plants (*Musa* AAB cv. Harton) and its relationship with physical, chemical and biological characteristics of the soil. Agron Costarricense 45(2):115–134. https://doi.org/10.15517/rac.v45i2.47772

González García HG, González AF, Pineda M, Escalante H, Yzquierdo GAR, Bracho AS (2021c) Edaphic microbiota in plantain lots of contrasting vigor and relationships with soil properties. Bioagro 33(2):143–148. https://doi.org/10.51372/bioagro332.8

González-Pedraza AF, Escalante J (2021) Nitrogen mineralization in soils cultivated with plantain (*Musa* AAB subgroup plátano cv. Hartón), Zulia state, Venezuela. Rev Fac Agron (LUZ) 38(3):525–547. https://n9.cl/so6u. Accessed 19 July 2021

González-Pedraza AF, Atencio J, Cubillán K, Almendrales R, Ramírez L, Barrios O (2014) Microbial activity in soils cultivated with plantain (Musa AAB plantain subgroup cv. Harton) with different vigor of plants. Rev Fac Agron (LUZ) 31:526–538. https://doi.org/10.51372/bioagro332.8

Guimarães RM, Lamandé M, Munkholm LJ, Ball BC, Keller T (2017) Opportunities and future directions for visual soil evaluation methods in soil structure research. Soil Tillage Res 173:104–113. https://doi.org/10.1016/j.still.2017.01.016

Hernández Y, Marín M, García J (2007) Response to the plantain (*Musa* AAB cv. Horn) yield as a function of the mineral nutrients and its phenologycal cycle. Part I. Growth and production. Rev Fac Agron (LUZ) 24:607–626. https://n9.cl/uwuxm. Accessed 19 Feb 2021

Hernández A, Bojorquez J, Morell F, Cabrera A, Ascanio M, García J, Madueño A, Nájera O (2010) Fundamentos de la estructura de suelos tropicales. Universidad Autónoma de Nayarit, Mexico

Johannes A, Weisskopf P, Schulin R, Boivin P (2017) To what extent do physical measurements match with visual evaluation of soil structure? Soil Tillage Res 173:24–32

Lal R (1998) Soil quality and agricultural sustainability. In: Lal R (ed) Soil quality and agricultural sustainability. Sleeping Bear Press, Inc, Chelsea, pp 3–12

Linting M, Meulman JJ, Groenen PJF, van der Kooij AJ (2007) Nonlinear principal components analysis: introduction and application. Psychol Methods 12:336–358. https://doi.org/10.1037/1082-989X.12.3.336

MacEwan RJ, Fitzpatrick RW (1996) The pedological context for assessment of soil quality. In: MacEwan RJ, Carter MR (eds), Soil quality is in the hands of the land manager. Proceedings of an international symposium. Advances in soil quality for land management: Science, Practice and Policy, 17–19 April 1996, University of Ballarat, Ballarat, Victoria, Australia, pp. 10–16

Martínez G, Delgado E, Rodríguez D, Hernández J, Del Valle R (2008) Brief analysis of Musaceae production in Venezuela. Prod Agrop 1:24–29. https://n9.cl/d48mt. Accessed 19 Feb 2021

Meulman JJ, van der Kooij AJ, Babinec A (2002) New features of categorical principal components analysis for complicated data sets, including data mining. In: Gaul W, Ritter G (eds) Classification, automation, and new media, pp 207–217. https://doi.org/10.1007/978-3-642-55991-4_22

Meulman JJ, van der Kooij AJ, Duisters KLW (2019) ROS regression: integrating regularization with optimal scaling regression. Stat Sci 34:361–390. https://doi.org/10.1214/19-STS697

Olivares BO, Araya-Alman M, Acevedo-Opazo C, Rey JC, Cañete P, Giannini F, Balzarini M, Lobo D, Navas-Cortés JA, Landa BB, Gómez JA (2020) Relationship between soil properties and banana productivity in the two main cultivation areas in Venezuela. J Soil Sci Plant Nutr 20:2512–2524. https://doi.org/10.1007/s42729-020-00317-8

Olivares BO, Calero J, Rey JC, Lobo D, Landa BB, Gómez JA (2022) Correlation of banana productivity levels and soil morphological properties using regularized optimal scaling regression. Catena 208:105718. https://doi.org/10.1016/j.catena.2021.105718

Pulido-Moncada M, Helwig Penning L, Timm LC, Gabriels D, Cornelis WM (2014) Visual examinations and soil physical and hydraulic properties for assessing soil structural quality of soils with contrasting textures and land uses. Soil Tillage Res 140:20–28. https://doi.org/10.1016/j.still.2014.02.009

Pulido-Moncada M, Penning LH, Timm LC, Gabriels D, Cornelis WM (2017) Visual examination of changes in soil structural quality due to land use. Soil Tillage Res 173:83–91. https://doi.org/10.1016/j.still.2016.08.011

Sevinç G, Aydogdu MH, Cancelik M, Sevinc MR (2019) Farmers' attitudes toward public support policy for sustainable agriculture in GAP-Sanliurfa, Turkey. Sustainability 11:6617. https://doi.org/10.3390/su11236617

Shepherd TG (2009) Visual soil assessment. In: Volume 1. Field guide for cropping and pastoral grazing on flat to rolling country, 2nd edn. Horizons Regional Council, Palmerston North

Soil Survey Staff (2014) Keys to soil taxonomy. Keys to soil taxonomy, 12th edn. USDA-Natural Resources Conservation Service, Washington, DC

Soil Survey Staff (2017) Soil survey manual handbook 18. United States Department of Agriculture, Washington, DC

Van der Kooij AJ (2007) Prediction accuracy and stability of regression with optimal scaling transformations. Unpublished doctoral dissertation, Leiden University, The Netherlands. https://n9.cl/ew8pz. Accessed 19 Feb 2021

Vanilarasu K, Balakrishnamurthy G (2014) Influences of organic manures and amendments in soil physiochemical properties and their impact on growth, yield and nutrient uptake of banana. Bioscan 9:525–529. https://n9.cl/awlm. Accessed 19 February 2021

Vaquero MR (2005) Soil physical properties and banana root growth. In: Turner DW, Rosales FE (eds) Banana root system: towards a better understanding for its productive management: proceedings of an international symposium held in San José, Costa Rica on 3–5 November 2003. International Network for the Improvement of Banana and Plantain, Montpellier, pp. 125–131. https://n9.cl/szt4m. Accessed 19 Feb 2021

Vasu D, Tiwari G, Sahoo S, Dash B, Jangir A, Sharma RP, Chandran P (2021) A minimum data set of soil morphological properties for quantifying soil quality in coastal agroecosystems. Catena 198(105042):1–12. https://doi.org/10.1016/j.catena.2020.105042

Vaughan IP, Ormerod SJ (2005) Increasing the value of principal components analysis for simplifying ecological data: a case study with rivers and river birds. J Appl Ecol 42:487–497. https://n9.cl/2wnp0. Accessed 19 Feb 2021

Villarreal-Núñez J, Pla-Sentis I, Agudo-Martínez L, Villaláz-Perez J, Rosales F, Pocasangre L (2013) Soil quality indexes in areas cultivated with bananas in Panama. Agron Mesoam 24:301–315. Available online at https://n9.cl/6dsn. Accessed 19 Feb 2021

Voorhees WB, Amemiya M, Allmaras RR, Larson WE (1971) Some effects of aggregate structure heterogeneity on root growth. Soil Sci Soc Am J 35(4):638–643. https://doi.org/10.2136/sssaj1971.03615995003500040043x

Chapter 7
Discussion and Conclusions

Abstract The general results were obtained through the integrated studies carried out on (a) a systematic review and bibliometric analysis of the agro-environmental factors that determine the appearance of *Fusarium oxysporum* f. sp. *cubense* Tropical Race 4 (*Foc* TR4) and the risk analysis and climatic suitability maps for syn. *Fusarium odoratissimum* in Venezuela through the maximum entropy model; (b) the relationship between soil properties and the incidence of banana wilt (MB) through Orthogonal Partial Least Squares Discriminant Analysis (OPLS-DA) and the random forest (RF) algorithm; (c) an analysis of the relationship between main physical, chemical, and biological soil properties most correlated to banana productivity; and (d) the potential use of soil morphological properties to differentiate banana productivity levels under Venezuelan conditions through a prediction model based on categorical soil properties with regularized optimal scaling regression. The results of the studies carried out show that science is progressing very quickly in various fields, such as epidemiology, phytopathology, soil microbiology, and soil morphology, among others, developing studies whose results can become tools that promote sustainable alternatives. Knowledge of the climatic suitability of *Foc* TR4, BW, the microbiome, and soil properties is essential to improve the productivity and sustainability of agricultural production. The combination of traditional disciplines with information technologies will surely revolutionize Venezuelan agriculture in the coming decades. The speed with which scientific results are brought to the market in some countries should be an example to follow for research groups in Venezuela. In conclusion, machine learning in agriculture could offer an advance that would guarantee decision-making with the objective of achieving improvements in banana productivity and plant health in Venezuela.

Keywords Artificial intelligence · Banana · Machine learning · Quality soil · Plant health · Sustainability

B. O. Olivares Campos, *Banana Production in Venezuela*, The Latin American Studies Book Series, https://doi.org/10.1007/978-3-031-34475-6_7

7.1 Introduction

The main challenges faced by the Venezuelan banana sector concerning the potential use of these machine learning algorithms are the training challenge, accessibility, ethics, and sustainability. The Venezuelan agri-food sector has to invest in training aimed at both commercial banana farms and communities of small owners. Facilitating access to this training that will result in greater food security for the entire territory should therefore be a priority on the agenda of the administrations (Olivares et al. 2016; Gründler and Krieger 2021).

About accessibility, network coverage must be guaranteed in rural environments in which the connectivity of the population has been neglected until now, enabling the functionality of artificial intelligence devices (Olivares et al. 2017). Regarding the ethical challenge, the correct and optimal use of machine learning algorithms must be able to guarantee food not only to the most developed countries, but it must also be a fundamental tool for the takeoff of other countries where food is still a scarce commodity. Finally, the sustainability factor (economic and environmental) must permeate this digitization and application of technologies to the banana sector, aligning with the sustainable development goals (Holloway and Mengersen 2018; Porciello et al. 2020).

Artificial intelligence, particularly machine learning, is a useful tool for analyzing amounts of data and providing complex nonlinear responses, thus advancing in the understanding of banana productivity processes and the design of useful strategies to combat the appearance of diseases in bananas. The complexity and multi-scale nature of the problem represented by the productivity and plant health of bananas make it a suitable field for the application of these techniques in different fields of study of the problem. However, its true usefulness emerges with the sharing of the parts studied by each specialty (e.g., edaphic studies, predictions of the risk of *Foc* TR4 appearance, among others).

The new approach that this book proposes about the potential influence of the soil on the productivity and plant health of bananas through the convergence of evidence is a good starting point for using varied and voluminous databases that are not unusual. They are often published in the best scientific journals. Through the different techniques based on artificial intelligence, responding to where and when the appearance of *Foc* TR4 and banana wilt could occur or which would be the most suitable areas to improve the productivity of banana plantations based on morphological properties of the soil becomes an exciting challenge. Knowing the impact of the biophysical and socioeconomic conditions of this serious problem is essential to design adequate solutions, which involve modifying or deactivating them. Interdisciplinary applications based on machine learning offer the opportunity to search for those paths of sustainability and provide relevant data and information for the management of resources that contribute to sustainable agriculture in territories such as those of Venezuela.

7.1.1 Dynamization of the Knowledge Transfer and Exchange

Soil diversity is ensuring primary production in terrestrial ecosystems and agricultural productivity (van Capelle et al. 2012). The cycle of water and nutrients, the formation and aggregation of soil (Olivares et al. 2022a, b), decomposition and carbon sequestration (Zhu et al. 2019), and the control of pest organisms and pathogens are essential functions in the soil that are driven by biota interactions (Hendrix et al. 2020). These functions support and regulate ecosystem services directed at agricultural production and sustainability in agricultural systems.

A primary objective of future banana-growing systems will be to maintain or increase agricultural productivity while maintaining sustainable production despite increasing demand for this fruit and continued soil degradation caused by inadequate soil management practices. Tillage, adequate fertilization, crop rotation, including cover crops, and agricultural residue management are the main tools that farmers, planners, environmental managers, and commercial farms use to design production soil management.

These soil management factors depend on farmers' decisions, triggered by regional traditions, local knowledge, and agricultural policies as impacts on governance. However, the initiative arises to reflect on the social role of universities and public research institutes in the knowledge society, which can be summarized in three key functions: (a) the transmission of knowledge through training activities (Olmos-Peñuela et al. 2014; Burmeister et al. 2020), (b) the generation of new knowledge through R&D activities (Owen et al. 2021), and (c) the transfer of knowledge and skills to the society that needs them, providing solutions to the specific problems of social and economic agents (Vega Jurado et al. 2011; Boenink and Kudina 2020).

Today, there is a considerable evolution in the relationship between scientific research organizations and social agents, which has made it possible to broaden the understanding of the processes of generation and application of knowledge. However, Venezuela is going through a process of economic, social, and political crisis that makes the relationship between science and society difficult. According to Castro Martínez and Vega Jurado (2009), this complexity is represented, among other factors, by the diversity of actors involved and the absence of clear borders that define their activities and functions: Companies carry out fundamental research activities, while research institutes' research is involved in processes of production and application of knowledge, creating, for example, companies and participating in entrepreneurial activities. Consequently, the relationship between the different actors is not unidirectional but is characterized by processes of collaboration and exchange, which has given rise to the concept of knowledge transfer and exchange (KTE) that is difficult to implement in the agri-food systems of Venezuela (Mitton et al. 2007; Castro Martínez and Vega Jurado 2009; Ribeiro et al. 2018).

This book provides insight to improve understanding of how the interaction of agriculture-based soil management practices affects biodiversity and how soil biodiversity in turn feeds back into soil functions and ecosystem services as factors for productivity and sustainability. For this, it is necessary to promote a cultural change in the scientific system and its relationship with innovation, contributing to increasing the involvement of research staff in the processes of exchange and transfer of knowledge with social agents in Venezuela.

Additionally, it is expected that the studies compiled in the book can be replicated in other areas of Venezuela, including Latin America and the Caribbean, so that in this way the interdependencies from the inputs of soil management through the habitat conditions and biodiversity to the sustainability and productivity of bananas (Martínez-Harms and Balvanera 2012; Potthoff et al. 2017), and that in turn allows identifying and developing best-practice soil management options that maximize biodiversity and conservation and sustainability by minimizing economic losses to farmers and society. This book is intended to favor Venezuelan agricultural research institutions for the implementation of active strategies and the promotion of KTE processes.

7.2 General Discussion

In this book, the relationships between soil, productivity, climatic suitability, and *Fusarium* wilt in banana have been studied, giving rise to several outcomes and results of scientific relevance and also, importantly, of practical implications for banana producers in Venezuela.

Based on the current knowledge of the epidemiology of *Foc*, the management practices oriented to increase the quality and suppressiveness of the soil are described in Chap. 3. These approaches can contribute to suppressing the inoculum of *Foc*, reduce the incidence and severity of the disease, and increase productivity. This will contribute to addressing the major challenge of identifying the most effective practices for decreasing the inoculum of a pathogen with long-term survival capacity that affects a perennial crop with continuous flowering and harvest cycles (Dita et al. 2018; Kema et al. 2021; García-Bastidas et al. 2020).

Secondly, an analysis of the climatic suitability of *Foc* RT4 for Venezuela-producing areas was performed in the study included in Chap. 3: an issue not accurately assessed before, despite the amount of information available on this pathogen. Our results showed how the maximum entropy model estimated that all the geographic areas described for banana production in Venezuela are climatically suitable for the occurrence of *Foc* RT4. Thus, in the current climate conditions, the areas of the great commercial importance of the municipalities La Cañada de Urdaneta and La Guajira in the Zulia state show the highest suitability scores (> 0.90), followed by the municipalities Buroz (0.80) and Brión (0.70) from the Miranda state, as well as the banana areas of the Trujillo state, specifically in the Bolívar and La Ceiba municipalities (0.80). These results are especially useful for the design of new strategies for the efficient use and greater effectiveness of the establishment of legislative and regulatory prevention activities on borders to avoid the introduction of *Foc* RT4 and or its early detection.

The intensification of research on *Foc* RT4 in Latin America and the Caribbean is extremely relevant, especially in aspects related to its epidemiology and control, particularly in Venezuela (Chap. 3). However, the unexpected recent detection of *Foc* RT4 in Colombia, Peru and Venezuela, with quite different producing situations in each of these countries, means that in all producing countries, resources must be dedicated to investigating the risks posed by *Foc* and on best management methods to address them (García-Bastidas et al. 2020). This chapter identified the most vulnerable areas in Venezuela (Zulia, Miranda, Trujillo, and Sucre states) for *Foc* RT4 occurrence according to a maximum entropy model and the main consequences for banana production due to a potential expansion in the geographical distribution of this pathogen in the future. This should help to develop the appropriate measures for control of the pathogen mainly based on early detection and eradication of *Foc* RT4 since no cure is available for this highly virulent race once is introduced in an area.

Chapter 4 focused on the analysis of the key soil properties that play an important role in the incidence of BW. So far, crop disease detection models primarily focus on leaf symptoms through image recognition technology. This means that diseases can be detected only after they have appeared. Our case study carried out using a machine learning technique for soils of lacustrine origin in Venezuela provided a complementary approach based on the relationship between BW and soil properties. This could be used to identify areas with different potential risks of high incidence of BW, optimizing the use of other techniques for BW detection and orienting best management practices to prevent its appearance. Our results showed that BW incidence is related to Zn, Fe, Ca, K, Mn, and clay content and provided a significant success (around 90%) in distinguishing between areas of high and low BW incidence. The soil properties identified are related to the drainage capacity of the soil (clay) or reflect indirectly the influence of different soil drainage conditions (e.g., Zn and Fe solubility).

Several studies pointed out that the determination of the soil properties in the banana zone studied in this thesis reveals similar information according to the reports by Rey et al. (2020) and Martínez et al. (2020). Machine learning analysis is considered an alternative to the more complex applications that deal with the prediction of banana diseases, as demonstrated by the performance of random forest in the investigations by Gómez-Selvaraj et al. (2020), Sangeetha et al. (2020), and Ye et al. (2020).

Our results open the field for further research in which we could quantitatively predict the risk of BW in banana fields based on available, or relatively easy to gather, information, which in turn could allow farm managers to implement preventive measures to minimize BW risk and target other techniques (e.g., plant sampling, withdrawal of infested material) to the areas where there is a maximum risk.

The way how soil properties determine banana productivity was quantified in Chap. 5. The penetration resistance, Mg, bulk density, and microbiological (biotic) properties such as microbial respiration and omnivorous free-living nematodes were linked to biometric and productive responses in commercial Cavendish banana plantations. This relationship with production can also be related to the response of plants to diseases. However, the variability of the systems must be considered in the eventual implementation of practices to manage soil properties for both production and disease management.

It was observed that in large banana farms (>50 ha), soil management is a critical factor that cannot be neglected since the productivity indicators evaluated in this study such as the number of hands per bunch and the circumference of the mother plant showed to have a close relationship with the physical properties: resistance to penetration and bulk density followed by the microbiological properties reported in recent studies of González-García et al. (2021a, b), Acevedo et al. (2021), and Olivares et al. (2022a, b), which establish that the deterioration of the soil could endanger the production of the crop in these soils.

The regression model obtained allowed the development of an easy-to-implement methodology for the rest of the banana areas of Venezuela and Latin America, which when further validated, would represent a tool for sustainable land management. When adapted to local considerations of each ecosystem, will allow monitoring and evaluating the quality of the soil and implementation actions that prevent the advance of degradation with a comprehensive and sustainable approach oriented also toward yield optimization.

Likewise, we consider that this study represents an important contribution to the knowledge of the banana soils of Venezuela, contributing to other recent studies focused on the quality of banana soil (González-García et al. 2021a, b; Rondón et al. 2021). We are aware that the study can be improved through its systematic use in new locations, so it is our intention and we hope that other research groups from the international scientific community will join in this task to produce improved versions of it.

In addition to corroborating the importance of individual soil properties and their interactions, Chap. 6 showed that it is possible to also consider soil morphological properties in the differentiation of banana productivity. There is little history of soil quality indices using categorical soil properties, and our study is a practical example to consider field morphological variables transformed by the optimal scale to capture the differences in banana productivity among sites.

The results of the study in Chap. 6 emphasize the importance of integrating morphological data in the numerical evaluation of soil quality. Our study is perhaps the first to present a categorical regression model of soil morphological properties to evaluate banana productivity. Therefore, the studied morphological indicators could be used effectively to assess soil quality and productivity in banana cropping systems as it has been made previously only for soil quality purposes, e.g., Pulido-Moncada et al. (2014) and Vasu et al. (2021).

Our results also showed that the categorical regression model with morphological properties is a potential option to quickly quantify the effect of land use type on soil quality, as it explains a system variability like that of other studies based on physical, chemical, and biological properties (Delgado et al. 2010a, b), the latter ones usually not feasible to be measured in a large number routinely for soil monitoring. So, the pedo-morphological properties selected in this study could serve as surrogates for other properties. Banana farmers in Venezuela can easily implement morphological soil quality assessment for rapid assessment of soil quality and link it to productivity. The approach we propose in this study reduces the time required for field evaluation of soil quality in banana-growing regions and, in any region, where soil data is limited.

Using these morphological properties can promote the participation of banana growers in the appraisal of soil quality and its link with banana productivity and so spur the interest of banana farmers in the introduction of best management practices. Although the participation of farmers in the visual evaluation of soil quality is sometimes limited and restricted by the lack of knowledge of pedo-morphology, current research shows the potential for addressing this limitation using digital soil morphometry with proximal sensors (Vasu et al. 2021; Olivares 2022). Looking to the future, the quantitative transformation of the morphological properties of the field soil by categorical regression will be useful in comparing the potential productivity of crops under different conditions.

Finally, most of the techniques developed in this book show the way to develop practical strategies for agricultural extension agents, providing a way to evaluate risk regarding banana diseases by agricultural services based on soil and climatic conditions to assist producers. This could also be used by producers and technical personnel of large- and medium-sized farms in banana production systems in Latin America and the Caribbean.

7.3 Key Conclusions

In summary, this book has identified the relationships between soil properties, productivity, climatic suitability, and the incidence of diseases in bananas in Venezuela. Compared to the previous studies, the most interesting and novel findings were:

(a) The approach of a systematic study that included a literature review, field observation, and systematic soil sampling on farms was effective in analyzing the role of soil in banana productivity and *Fusarium* wilt incidence.
(b) A review of the current state of knowledge concerning *Foc* TR4 worldwide and an analysis of the environmental suitability in Venezuela were provided, which made it possible to understand the real risk posed by *Foc* TR4 at the national and global levels.
(c) The presence-only data of the *Foc* pathogen on a world scale helped to model the future distributions of this species in Venezuela and to inform about the approximate risk of appearance and transmission of this lethal disease in bananas, indicating that the banana areas with greater commercial importance in Venezuela present adequate climatic conditions for survival and possibly the establishment of *Foc* TR4 in case of its introduction.
(d) The potential of using soil characteristics (Zn, Fe, K, Ca, Mn, and clay content) as predictors to avoid the establishment of banana plantations in soils with a high risk of incidence of banana wilt disease has been demonstrated using the random forest algorithm.

(e) The influence of specific soil properties such as Mg, penetration resistance, total microbial respiration, bulk density, and omnivorous free-living nematodes on banana productivity was demonstrated, with the interactions among them appearing to be more significant than the effect of a single property on the productivity.
(f) The soil type (lacustrine and alluvial) can naturally affect banana productivity so that each type of soil should be considered individually when identifying the soil properties that can explain differences in banana productivity.
(g) The categorical regression analysis generated an operational model in bananas in which the soil morphological properties, such as biological activity, texture, dry consistence, reaction to HCl, and structure type, differentiated different banana productivity levels. The results of this analysis showed for the first time that soil morphological variables could be promising new soil indicators for the assessment of banana productivity in Venezuelan soils.

References

Acevedo I, Sánchez A, Mendoza B (2021) Evaluación del nivel de degradación del suelo en dos sistemas productivos en la depresión de Quíbor. II. Calidad del suelo. Bioagro 33(2):127–134. https://doi.org/10.51372/bioagro332.6

Boenink M, Kudina O (2020) Values in responsible research and innovation: from entities to practices. J Responsible Innov 7(3):450–470

Burmeister A, Li Y, Wang M, Shi J, Jin Y (2020) Team knowledge exchange: how and when does transformational leadership have an effect? J Organ Behav 41(1):7–31

Castro Martínez E, Vega Jurado J (2009) Las relaciones universidad-entorno socioeconómico en el Espacio Iberoamericano del Conocimiento. Rev iberoam cienc tecnol soc 4(12):71–81

Delgado E, Trejos J, Villalobos M, Martínez G, Lobo D, Rey J, Rodríguez G, Rosales F, Pocasangre L (2010a) Determinación de un índice de calidad y salud de suelos para plantaciones bananeras en Venezuela. Interciencia 35(12):927–933. https://n9.cl/nv7vv

Delgado E, Rosales F, Trejos J, Villalobos M, Pocasangre L (2010b) Índice de calidad y salud de suelos para plantaciones bananeras de cuatro países de América Latina y el Caribe. Bioagro 22(1):53–60. https://n9.cl/qm3e

Dita M, Barquero M, Heck D, Mizubuti ESG, Staver CP (2018) Fusarium wilt of banana: current knowledge on epidemiology and research needs toward sustainable disease management. Front Plant Sci 9(1468):1–21. https://doi.org/10.3389/fpls.2018.01468

García-Bastidas FA, Quintero-Vargas JC, Ayala-Vasquez M, Schermer T, Seidl MF, Santos-Paiva M et al (2020) First report of Fusarium wilt tropical race 4 in Cavendish bananas caused by *Fusarium odoratissimum* in Colombia. Plant Dis 104:994–994. https://doi.org/10.1094/PDIS-09-19-1922-PDN

Gómez-Selvaraj M, Vergara A, Montenegro F, Alonso Ruiz H, Safari N, Raymaekers D, Ocimati W, Ntamwira J, Tits L, Omondi AB, Blomme G (2020) Detection of banana plants and their major diseases through aerial images and machine learning methods: a case study in DR Congo and Republic of Benin. ISPRS 169:110–124. https://doi.org/10.1016/j.isprsjprs.2020.08.025

González-García H, González Pedraza AF, Atencio J, Soto A (2021a) Evaluación de calidad de suelos plataneros a través de la actividad microbiana en el sur del lago de Maracaibo, estado de Zulia, Venezuela. Rev Fac Agr (LUZ) 38(2):216–240. https://doi.org/10.47280/RevFacAgron(LUZ).v38.n2.01

González-García H, González Pedraza AF, Pineda M, Escalante-García H, Rodríguez-Yzquierdo G, Soto-Bracho A (2021b) Microbiota edáfica en lotes de plátano con vigor contrastante y su relación con propiedades del suelo. Bioagro 33(2):143–148. https://doi.org/10.51372/bioagro332.8

Gründler K, Krieger T (2021) Using machine learning for measuring democracy: an update. ZEW-CEPR discussion paper (21-012). https://doi.org/10.2139/ssrn.3792811

Hendrix PF, Crossley DA, Blair JM, Coleman DC (2020) Soil biota as components of sustainable agroecosystems. In: Sustainable agricultural systems. CRC Press, Boca Raton, pp 637–654

Holloway J, Mengersen K (2018) Statistical machine learning methods and remote sensing for sustainable development goals: a review. Remote Sens 10(9):1365. https://doi.org/10.3390/rs10091365

Kema GHJ, Drenth A, Dita M, Jansen K, Vellema S, Stoorvogel JJ (2021) Editorial: fusarium wilt of banana, a recurring threat to global banana production. Front Plant Sci 11:628888. https://doi.org/10.3389/fpls.2020.628888

Martínez G, Rey-Brina JC, Pargas-Pichardo RE, Domínguez M (2020) Actualización sobre el manejo de la marchitez por Fusarium Raza 4 Tropical en musáceas. Agron Trop 70:1–18. https://doi.org/10.5281/zenodo.4291108

Martínez-Harms MJ, Balvanera P (2012) Methods for mapping ecosystem service supply: a review. Int J Biodivers Sci Ecosyst Serv Manag 8(1–2):17–25

Mitton C, Adair CE, McKenzie E, Patten SB, Perry BW (2007) Knowledge transfer and exchange: review and synthesis of the literature. Milbank Q 85(4):729–768

Olivares BO (2022) Machine learning and the new sustainable agriculture: applications in banana production systems of Venezuela. Agric Res Updates 42:133–157

Olivares BO, Cortez A, Muñetones A, Casana S (2016) Strategic elements of organizational knowledge management for innovation. Case: agrometeorology network. RIDU 10(1):68–81. https://doi.org/10.19083/ridu.10.446

Olivares BO, Lobo D, Cortez A, Rodríguez MF, Rey JC (2017) Socio-economic characteristics and methods of agricultural production of indigenous community Kashaama, Anzoategui, Venezuela. Rev Fac Agron (LUZ) 34(2):187–215. https://n9.cl/p2gc5

Olivares BO, Calero J, Rey JC, Lobo D, Landa BB, Gómez JA (2022a) Correlation of banana productivity levels and soil morphological properties using regularized optimal scaling regression. Catena 208:105718

Olivares BO, Rey JC, Perichi G, Lobo D (2022b) Relationship of microbial activity with soil properties in banana plantations in Venezuela. Sustainability 14:13531. https://doi.org/10.3390/su142013531

Olmos-Peñuela J, Castro-Martínez E, d'Este P (2014) Knowledge transfer activities in social sciences and humanities: explaining the interactions of research groups with non-academic agents. Res Policy 43(4):696–706

Owen R, von Schomberg R, Macnaghten P (2021) An unfinished journey? Reflections on a decade of responsible research and innovation. J Responsible Innov 8(2):217–233

Porciello J, Ivanina M, Islam M, Einarson S, Hirsh H (2020) Accelerating evidence-informed decision-making for the Sustainable Development Goals using machine learning. Nat Mach Intell 2(10):559–565. https://doi.org/10.1038/s42256-020-00235-5

Potthoff M, Pérès G, Taylor A, Schrader S, Landa B, Nicolai A, Sandor M, Öptik M, Gema G, Bergmann H, Cluzeau D (2017) Ecosystem services driven by the diversity of soil biota-understanding and management in agriculture-the Biodiversa SoilMan-project. In EGU general assembly conference abstracts 2017 Apr (p. 5356)

Pulido-Moncada M, Gabriels D, Lobo D, Rey BJC, Cornelis WM (2014) Visual field assessment of soil structural quality in tropical soils. Soil Tillage Res 139:8–18. https://doi.org/10.1016/j.still.2014.01.002

Rey JC, Martínez-Solórzano G, Ramírez H, Pargas-Pichardo R (2020) Relación de las condiciones agroecológicas de un lote de planicie lacustrina con la marchitez del banano Cavendish en Aragua, Venezuela. Agron Trop 70:1–12. https://doi.org/10.5281/zenodo.4346252

Ribeiro B, Bengtsson L, Benneworth P, Bührer S, Castro-Martínez E, Hansen M, Shapira P (2018) Introducing the dilemma of societal alignment for inclusive and responsible research and innovation. J Responsible Innov 5(3):316–331

Rondón T, Hernandez RM, Guzman M (2021) Soil organic carbon, physical fractions of the macro-organic matter, and soil stability relationship in lacustrine soils under banana crop. PLoS One 16(7):e0254121. https://doi.org/10.1371/journal.pone.0254121

Sangeetha T, Lavanya G, Jeyabharathi D, Rajesh Kumar T, Mythili K (2020) Detection of pest and disease in the banana leaf using convolution random forest. Test Eng Manag 83(2):3727–3735. https://n9.cl/yeffq

van Capelle C, Schrader S, Brunotte J (2012) Tillage-induced changes in the functional diversity of soil biota – a review with a focus on German data. Eur J Soil Biol 50:165–181

Vasu D, Tiwari G, Sahoo S, Dash B, Jangir A, Sharma RP, Chandran P (2021) A minimum data set of soil morphological properties for quantifying soil quality in coastal agroecosystems. Catena 198(105042):1–12. https://doi.org/10.1016/j.catena.2020.105042

Vega Jurado JM, Manjarrés Henríquez LA, Castro-Martínez E, Fernández de Lucio I (2011) Las relaciones universidad-empresa: tendencias y desafíos en el marco del Espacio Iberoamericano del Conocimiento. Rev Iberoam de Educ 57:109–124

Ye HC, Huang WJ, Huang SY, Cui B, Dong YY, Guo AT, Ren Y, Jin Y (2020) Identification of banana fusarium wilt using supervised classification algorithms with UAV-based multi-spectral imagery. Int J Agric Biol Eng 13(3):136–142. https://doi.org/10.25165/j.ijabe.20201303.5524

Zhu YG, Zhao Y, Zhu D, Gillings M, Penuelas J, Ok YS, Banwart S (2019) Soil biota, antimicrobial resistance and planetary health. Environ Int 131:105059

Index

B. O. Olivares Campos, *Banana Production in Venezuela*, The Latin American Studies Book Series, https://doi.org/10.1007/978-3-031-34475-6

MIX
Papier aus verantwortungsvollen Quellen
Paper from responsible sources
FSC® C105338

If you have any concerns about our products,
you can contact us on
ProductSafety@springernature.com

In case Publisher is established outside the EU,
the EU authorized representative is:
Springer Nature Customer Service Center GmbH
Europaplatz 3, 69115 Heidelberg, Germany

Printed by Libri Plureos GmbH
in Hamburg, Germany